SOCIAL MEDIA MARKETING & PERSONAL BRANDING

A STEP-BY-STEP SYSTEM TO BUILD AUTHORITY, ATTRACT FOLLOWERS, AND MONETIZE WITH AI AND SEO ACROSS TIKTOK, YOUTUBE, LINKEDI, INSTAGRAM, FACEBOOK, AND X

DAVE STONE

TOP TIER PRESS

Published under the author brand **Dave Stone**

This edition contains material originally published separately as:

Book One: ***Social Media Marketing* © 2025 Top Tier Press**

Book Two: ***Personal Branding* © 2025 Top Tier Press**

Contents

Part I
Social Media Marketing

Part II
Personal Branding

PART I
Social Media Marketing

CRACK THE CODE TO ONLINE SUCCESS WITH AI-POWERED CONTENT, SEO MASTERY, AND PROVEN STRATEGIES TO MONETIZE TIKTOK, INSTAGRAM, FACEBOOK, YOUTUBE, AND BUILD PASSIVE INCOME STREAMS

Introduction

The landscape of making money has fundamentally shifted. While traditional careers still exist, a parallel economy has emerged where teenagers become millionaires through TikTok dances, home cooks build six-figure businesses through Instagram Reels, and everyday people generate passive income that rivals corporate salaries, all from their smartphones.

This is not about luck or going viral once. It is about understanding the systems, algorithms, and psychology that drive social media success. More importantly, it is about leveraging artificial intelligence and search engine optimization to work smarter, not harder.

WHY SOCIAL MEDIA IS THE MOST POWERFUL INCOME OPPORTUNITY TODAY

Consider this: traditional businesses need storefronts, inventory, employees, and significant capital. Social media businesses need creativity, consistency, and strategy. The barrier to entry is almost nonexistent, but the ceiling for success is limitless.

Every scroll, like, and share represents a micro-transaction of attention. Platforms such as TikTok, Instagram, Facebook, and YouTube have become sophisticated marketplaces where attention converts into influence, and influence converts into income. The companies that own these platforms have already proven this model works. They have built trillion-dollar valuations from it.

HOW AI, SEO, AND PLATFORM STRATEGIES INTERSECT

The game has evolved beyond posting random content and hoping for the best. Successful creators now understand that artificial intelligence can generate content ideas, optimize posting schedules, and even write captions that perform better than anything they could create manually. They know that search engine optimization principles apply to social media algorithms, making their content discoverable not only today but for years to come.

This intersection creates a compound effect. AI helps you create more content faster. SEO principles help that content get discovered consistently. Platform-specific strategies help it convert into actual income. When these three elements work together, the results are exponential rather than linear.

WHAT THIS BOOK WILL TEACH YOU

You are about to learn the exact systems that turn social media from a time-wasting distraction into a wealth-building machine. This is not theory or motivational talk. Every strategy has been tested, refined, and proven to work across multiple platforms and niches.

By the final chapter, you will have a complete toolkit that includes AI prompts for content creation, SEO formulas for maximum reach, platform-specific growth strategies, and multiple monetization methods that generate income while you sleep.

SETTING EXPECTATIONS: FROM OVERWHELMED BEGINNER TO CONFIDENT ONLINE ENTREPRENEUR

Right now, you might feel overwhelmed by the constant changes in social media. New features launch daily, algorithms shift without warning, and trends die as quickly as they appear. This chaos is exactly why systematic approaches work better than random tactics.

Think of this book as your GPS through the social media landscape. You do not need to understand every street or shortcut; you just need to know where you are going and trust the directions. By following the roadmap laid out in these chapters, you will build something sustainable that adapts to changes instead of breaking with every algorithm update.

The transformation from consumer to creator to entrepreneur happens in stages. You will start by understanding why social media works as a business model, then learn to use AI and SEO as powerful tools, master individual platforms, create content that converts, and finally build systems that generate passive income.

Your relationship with social media is about to change forever. Instead of scrolling mindlessly, you will see opportunities. Instead of consuming content, you will create value. Instead of spending money on these platforms, you will make money from them.

The new age of social media wealth is not coming. It is already here. Your position in this economy is entirely up to you.

Unlock Your Social Authority Playbook

Ready to turn your online presence into lasting influence and income?

You've mastered the basics. Now it's time to go beyond. **The Social Authority Playbook** gives you clear systems and tools to scale your brand, earn more, and grow with confidence. Inside, you'll find proven tactics to automate your content, expand your reach, and build genuine trust that lasts.

What you'll discover inside::

- Build loyal digital communities
- Streamline your content systems
- Monetize your personal expertise
- Future proof your online brand

☞ **Access Your Playbook Now**

Visit: **https://content.toptierpress.com/amplify-online-influence**

Or scan the QR code below to get instant access!

ONE

Cracking the Code

WHY SOCIAL MEDIA IS THE NEW ECONOMY

Social media platforms didn't start as business engines, but they've evolved into the most sophisticated commerce systems ever created. Facebook began as a college directory, Instagram started as a photo-sharing app, and TikTok emerged as a lip-syncing platform. Today, these platforms facilitate billions in transactions and have created more millionaires under 30 than any industry in history.

The transformation happened because these platforms solved the biggest challenge in business: finding customers. Traditional businesses spend enormous resources on advertising, market research, and customer acquisition. Social media platforms provide direct access to billions of people, complete with detailed demographic information, behavioral data, and real-time feedback mechanisms.

Consider how Amazon revolutionized retail by removing friction from purchasing. Social media platforms have done something even more powerful—they've removed friction from building relationships with potential customers. You can now test product ideas, validate market demand, build anticipation, and process sales without ever leaving the platform ecosystem.

The numbers support this shift. E-commerce sales driven by social media exceeded $700 billion globally in 2023, and that figure grows exponentially each year. More telling is the speed of growth. Traditional businesses might take years to reach their first million in revenue. Social media businesses can achieve this in months or even weeks.

Platform Economics 101

Every social media platform operates on the same basic economic model: they capture attention, analyze behavior, and sell access to that attention. As a creator or business owner, you're essentially partnering with these platforms. They provide the audience and infrastructure; you provide the content and value.

This partnership model explains why platforms actively support successful creators through monetization programs, algorithm boosts, and promotional features. When you succeed, they succeed. Understanding this alignment is crucial because it means working with the platform's objectives rather than against them.

The most successful social media entrepreneurs think like platform partners, not just users. They understand that platforms reward behavior that keeps users engaged, attracts new users, and drives on-platform transactions. This knowledge allows them to create content that naturally aligns with algorithmic preferences.

THE PSYCHOLOGY OF ATTENTION & INFLUENCE

Attention is the new currency, and social media platforms are the banks. But attention alone isn't valuable; conversion is. The ability to turn attention into influence and influence into income requires understanding fundamental psychological principles that drive human behavior online.

The psychology of social media engagement operates on several key principles. First, social proof: people follow crowd behavior and trust content that others have validated through likes, shares, and comments. Second, reciprocity: audiences engage with creators who provide value before asking for anything in return. Third, authority: people trust creators who demonstrate expertise and authenticity in their niche.

These psychological triggers aren't manipulative tactics; they're communication principles that help you build genuine connections with your audience. The most successful creators understand that social media success comes from serving others' needs, not just promoting your own agenda.

The Influence Equation

Influence on social media follows a predictable equation: Consistency + Value + Authenticity = Influence. Consistency builds recognition and trust. Value ensures people have reasons to follow and engage. Authenticity differentiates you from competitors and creates emotional connections.

This equation explains why buying followers or engagement doesn't work long-term. Purchased metrics lack the psychological validation of genuine engagement. Algorithms can detect artificial engagement patterns, and real audiences can sense inauthentic content.

Building genuine influence takes time, but the results compound exponentially. Each piece of valuable content you create has the potential to attract new followers who become customers, collaborators, or advocates for your brand.

The Attention Economy Hierarchy

Not all attention is equal. Social media operates on a hierarchy where different types of engagement carry different values. At the bottom are passive views and follows. In the middle are active engagements like comments, shares, and saves. At the top are conversions, actions that move people from social media to your owned platforms like email lists, websites, or direct purchases.

Understanding this hierarchy helps you create content strategies that move people up the engagement ladder. You can't start by asking for sales; you need to earn attention, build trust, and provide value first. But once you understand the progression, you can design content that naturally guides people toward higher-value actions.

WHY NOW IS THE BEST TIME TO BUILD YOUR BRAND

The social media landscape has matured to the point where individual creators can compete with major brands, but it hasn't become so saturated that new entrants can't succeed. This represents a unique window of opportunity that combines accessibility with profitability.

Several factors make this the optimal time to build your social media presence. First, AI tools have democratized high-quality content creation. You no longer need expensive equipment, design skills, or large teams to produce professional content. Second, platform monetization has evolved beyond advertising to include direct payments, e-commerce integration, and creator funds.

Third, audiences have become more sophisticated in their content preferences but also more willing to support creators they value. The stigma around online purchases has disappeared, and digital products are now considered as valuable as physical ones.

Market Timing Advantages

Early adopters of social media monetization faced significant challenges. Payment systems were clunky, audience trust was low, and platform support was minimal. Today's creators benefit from refined systems, established user behaviors, and platform-provided tools that handle much of the technical complexity.

The current landscape rewards creators who focus on value creation rather than growth hacking. This shift favors sustainable business models over viral gimmicks. While viral content can still provide massive exposure, consistent value creation builds lasting businesses.

Additionally, the economic climate has accelerated the shift toward digital entrepreneurship. Traditional job security has decreased while online income opportunities have increased. This creates both necessity and opportunity for individuals to build alternative income streams through social media.

The Compound Effect of Starting Now

Social media success compounds over time. Content you create today can continue attracting audiences and generating income for years. Links build authority, followers become customers, and customers become advocates who attract more followers.

This compound effect means that starting today, even with modest results, positions you ahead of someone who starts next year with the same effort. The platforms, tools, and strategies available now are more sophisticated than ever, but they're still accessible enough that dedication and smart strategy can overcome larger budgets or established competition.

The creators who will dominate the next decade are those who start building systematic, value-driven social media businesses today. The tools are ready, the audiences are waiting, and the economic incentives are aligned. The only question is whether you'll take advantage of this opportunity.

Action Steps for Chapter 1:

Action Item	Task Description	Expected Outcome
Platform Audit	Choose one primary platform where your target audience is most active. Research the top creators in your niche and analyze their content patterns, posting frequency, and engagement strategies.	Clear understanding of your optimal platform and competitive landscape
Value Identification	Write down three specific problems you can solve for your potential audience. These will become the foundation for your content strategy.	Defined value proposition and content pillars
Competition Analysis	Identify five creators in your space who have the audience size you want to achieve. Study their content for 30 days, noting what generates the highest engagement and what monetization methods they use.	Strategic insights into proven success patterns
Influence Baseline	Create your first piece of valuable content using the Consistency + Value + Authenticity formula. This establishes your starting point for measuring growth.	First published content and baseline metrics
Long-term Vision	Define what success looks like for your social media business in 12 months. Set specific metrics for followers, engagement rate, and income goals.	Clear 12-month roadmap with measurable objectives

TWO

AI as Your Secret Weapon

Artificial intelligence has fundamentally changed the content creation game, but most creators are still using it wrong. They're asking AI to write their captions or generate ideas, treating it like a more sophisticated Google search. The real power of AI lies in using it as a creative partner that amplifies your unique voice and systematizes your content production.

The transformation begins with understanding that AI doesn't replace creativity. It removes the friction from expressing it. Instead of staring at a blank screen wondering what to post, you can feed AI your expertise, audience insights, and brand voice to generate content frameworks that you then personalize and refine.

Consider the traditional content creation process: brainstorming ideas, researching topics, writing copy, editing for clarity, optimizing for platforms, and scheduling posts. This workflow could easily consume 3-4 hours per piece of content. AI can compress this timeline to 30-45 minutes while often improving the quality of the final output.

The AI Content Multiplication Effect

The most powerful application of AI in content creation is multiplication, not replacement. Start with one core idea, insight, or experience from your expertise. AI can help you transform this single concept into multiple content formats across different platforms.

Take a business tip about productivity. AI can help you turn this into a Twitter thread, an Instagram carousel, a TikTok script, a LinkedIn article, and a YouTube video outline. Each piece maintains your core message but adapts to platform-specific audiences and formats. This approach ensures consistency across your brand while maximizing the value extracted from each insight.

The key is training AI to understand your voice, values, and audience preferences. This means feeding it examples of your best-performing content, your brand guidelines, and detailed audience personas. The more context you provide, the more sophisticated and on-brand the outputs become.

Essential AI Tools for Content Creators

Different AI tools excel at different aspects of content creation. ChatGPT and Claude are excellent for ideation, copywriting, and strategic planning. Midjourney and DALL-E create visual content that would otherwise require design skills or expensive stock photo subscriptions. Jasper and Copy.ai specialize in marketing copy that converts.

But the real magic happens when you combine tools systematically. Use ChatGPT to generate content ideas based on trending topics in your niche. Feed those ideas to Midjourney for accompanying visuals. Use Grammarly's AI features to polish the copy, and tools like Buffer or Hootsuite to optimize posting schedules based on audience behavior data.

The goal isn't to become dependent on AI tools, but to create workflows where AI handles routine tasks so you can focus on strategy, relationship building, and high-level creative decisions. Think of AI as your content production team, handling research, first drafts, and optimization while you provide vision, expertise, and final approval.

Prompt Engineering for Content Success

The difference between mediocre and exceptional AI outputs lies in prompt engineering. The art and science of communicating effectively with AI systems. Generic prompts produce generic content. Specific, contextual prompts generate content that feels human-created and audience-focused.

Instead of asking, "Write an Instagram caption about productivity," try: "Write an Instagram caption for busy entrepreneurs who struggle with time management. The tone should be empathetic but motivational. Include a personal anecdote about overcoming procrastination and three actionable tips they can implement today, and end with a question that encourages comments about their biggest time management challenges."

This approach transforms AI from a content generator into a strategic partner that understands your audience, objectives, and brand voice. The more detailed and specific your prompts, the less editing and personalization required in the final output.

CREATING CONSISTENCY WITHOUT BURNOUT

Consistency is the foundation of social media success, but it's also the primary cause of creator burnout. The pressure to produce fresh, engaging content daily leads many creators to abandon their efforts within months. AI solves this problem by creating sustainable systems for consistent content production without creative exhaustion.

The secret lies in understanding that consistency doesn't mean constant creation—it means systematic production. Instead of creating content reactively when you feel inspired, AI enables proactive content creation where you batch produce weeks or months of content during peak creative periods.

This systematic approach transforms content creation from a daily stress into a strategic business process. You can dedicate specific days to content creation, knowing that AI will help you maximize output during those focused sessions. The rest of your time can be spent on engagement, strategy, and business development.

The Content Banking System

Content banking involves creating and storing content in advance, building a reserve that ensures consistent posting even during busy periods. AI makes this process exponentially more efficient by helping you generate multiple variations of successful content themes.

Start by identifying your 10-15 most successful pieces of content. Use AI to analyze what made them successful: the topics, formats, emotional triggers, and audience responses. Then systematically create variations of these successful frameworks, adapting them for different platforms and current events.

For example, if a "behind-the-scenes" post performed well, use AI to generate 20 different behind-the-scenes content ideas that reveal various aspects of your work or creative process. Bank these ideas with accompanying captions and posting schedules. Instead of creating content from scratch, you can draw from a well-researched and pre-approved collection when you need it.

Automation That Preserves Authenticity

The biggest concern with AI-assisted content is losing authenticity. Audiences can detect overly automated or generic content, and platforms increasingly favor authentic engagement over polished but impersonal posts. The solution is strategic automation that preserves your unique voice while streamlining production.

Use AI for ideation, first drafts, and optimization, but maintain human oversight for final approval, personalization, and community interaction. Automate the mechanics of content creation, research, formatting, scheduling, while keeping the strategic decisions and audience engagement human-driven.

This hybrid approach allows you to maintain authentic connections with your audience while producing content at a scale that would be impossible through manual processes alone. Your personality, expertise, and values remain central to the content, but AI handles the time-consuming production tasks.

The Energy Management Framework

Sustainable content creation requires managing creative energy as carefully as managing time. AI helps optimize your energy by handling routine tasks during low-energy periods and maximizing creative output during peak performance times.

Track your natural energy patterns and align content creation tasks accordingly. Use high-energy periods for strategic planning, original content creation, and audience engagement. Use low-energy periods for AI-assisted tasks like research, editing, and scheduling.

This framework prevents the creative burnout that destroys many promising social media careers. By working with your natural rhythms and using AI to fill the gaps, you can maintain consistency without sacrificing quality or personal well-being.

CASE STUDIES: AI-ASSISTED VIRAL CONTENT

Real-world success stories demonstrate how AI can amplify human creativity to achieve viral reach and significant business results. These case studies reveal specific strategies and tools that transformed ordinary content into extraordinary performance.

Case Study 1: The Educational Carousel Revolution

A financial literacy educator struggled to make complex topics engaging on Instagram. Using AI, she developed a system that transformed dense financial concepts into visually appealing carousel posts that went viral consistently.

Her process began with AI research to identify the most searched financial questions among her target demographic. She then used ChatGPT to break complex concepts into simple, actionable steps. AI tools helped her create consistent visual templates and optimize captions for maximum engagement.

The breakthrough came when she used AI to analyze her top-performing posts and identify patterns in language, structure, and visual elements that resonated with her audience. She created templates based on these insights, allowing her to produce high-quality educational content systematically.

Results: Her follower count grew from 15,000 to 250,000 in eight months, and she launched a financial coaching program that generated six figures in its first year. The key was using AI not to replace her expertise, but to make it more accessible and engaging.

Case Study 2: The Trending Topic Amplification

A lifestyle blogger used AI to capitalize on trending topics without losing her authentic voice. Instead of jumping on trends reactively, she created a system that anticipated trends and prepared content in advance.

Her AI workflow monitored social media platforms for emerging trends in her niche, analyzed which trends aligned with her brand values, and generated content ideas that connected trends to her expertise. This proactive approach allowed her to be among the first creators to produce high-quality content on trending topics.

The system included AI-generated backup content for different types of trends, seasonal, news-related, cultural moments, and industry developments. When trends emerged, she could quickly adapt her prepared frameworks rather than starting from scratch.

Results: Her average post engagement increased by 340%, and she secured three major brand partnerships within six months. The combination of timeliness and quality, enabled by AI preparation, positioned her as a thought leader in her space.

Case Study 3: The Multi-Platform Optimization

A business coach struggled to maintain presence across multiple platforms while running his coaching business. AI helped him create a content multiplication system that turned single insights into platform-optimized content across TikTok, LinkedIn, Instagram, and YouTube.

He began by recording short voice memos about business ideas while he was on his way to work. AI transcribed these memos and transformed them into platform-specific content: TikTok scripts, LinkedIn articles, Instagram captions, and YouTube video outlines. Each piece maintained his core message while adapting to platform-specific audiences and formats.

AI also helped him optimize posting times, hashtags, and content formats for each platform based on performance data. The system allowed him to maintain consistent presence across four platforms while spending only 90 minutes per week on content creation.

Results: His combined follower count across platforms grew from 50,000 to 400,000 in one year, and his coaching business revenue increased by 250%. The efficiency gained through AI allowed him to focus on high-value activities like client work and business development.

Pattern Analysis: What Made These Cases Successful

These case studies reveal common patterns in successful AI-assisted content creation. First, AI was used to amplify existing expertise, not replace it. Each creator maintained their unique perspective and values while using AI to express them more effectively.

Second, successful AI implementation focused on systems and processes rather than one-off content creation. The creators developed repeatable workflows that consistently produced high-quality results rather than depending on occasional bursts of inspiration.

Third, they used AI for research and optimization, not just content generation. Understanding audience preferences, trending topics, and platform algorithms through AI analysis informed their content strategy and improved performance systematically.

Finally, they maintained the human element in final content approval and audience engagement. AI handled production efficiency, but human judgment determined strategic direction and authentic connection with audiences.

Action Steps for Chapter 2:

Action Item	Task Description	Expected Outcome
AI Tool Setup	Choose one primary AI tool (ChatGPT, Claude, or Jasper) and create detailed prompts for your niche. Include your brand voice, target audience, and content objectives in your prompt templates.	Personalized AI toolkit ready for consistent content creation
Content Banking Weekend	Dedicate 4-6 hours to creating 30 days of content ideas using AI. Generate topics, captions, and visual concepts that align with your expertise and audience needs.	30-day content calendar eliminating daily creation pressure

Action Item	Task Description	Expected Outcome
Workflow Optimization	Map out your current content creation process and identify three tasks that AI could handle or improve. Implement AI assistance for these tasks over the next week.	Streamlined content workflow with 50%+ time savings
Voice Training	Feed your chosen AI tool examples of your best-performing content to help it understand your style. Create a "brand voice document" that you can reference in future prompts.	AI system trained to produce on-brand content consistently
Performance Tracking	Establish baseline metrics for your current content performance. After implementing AI-assisted content creation for 30 days, measure improvements in engagement, reach, and time efficiency.	Data-driven proof of AI impact on content performance

THREE

The Foundations of Online Success

Your brand is not your logo, color scheme, or aesthetic, though those elements matter. Your brand is the emotional and intellectual connection people form with you and your content. It's the reason someone chooses to follow you instead of the thousands of other creators in your space, and why they eventually become paying customers rather than passive observers.

Building a strong personal brand on social media requires clarity in three fundamental areas: who you are (identity), how you communicate (voice), and why people should trust you (authority). Most creators skip the foundational work and jump straight to content creation, which explains why their growth stagnates after the initial excitement wears off.

Identity goes beyond demographics and interests. It encompasses your values, experiences, perspective, and the unique combination of skills and knowledge that only you possess. Your identity becomes your differentiator in crowded markets. When someone can get similar information from dozens of creators, they choose the one whose identity resonates most strongly with their aspirations or challenges.

Crafting Your Authentic Identity

Authenticity doesn't mean sharing everything about your life or being vulnerable for the sake of vulnerability. It means being genuinely yourself within the context of the value you provide. Your authentic identity is the intersection of who you truly are and what your audience needs.

Start by identifying the experiences, challenges, and victories that shaped your expertise. These stories become the foundation of your content because they're uniquely yours. A fitness coach who struggled with body image creates different content than one who was always athletic. A business mentor who failed multiple times before succeeding offers different insights than one who found success immediately.

Your identity should feel natural to maintain long-term. If you're introverted, don't build a brand that requires constant high-energy content. If you're naturally analytical, don't force yourself into a purely emotional content style. The goal is to amplify your natural strengths and characteristics, not create a persona you can't sustain.

Developing Your Brand Voice

Your brand voice is how your personality translates into written and spoken communication. It encompasses tone, vocabulary, humor level, formality, and the emotional atmosphere you create through your content. A well-defined brand voice makes your content instantly recognizable, even without seeing your name or profile picture.

Consider the difference between a brand voice that's "encouraging and practical" versus one that's "provocative and challenging." Both can be effective in the same niche, but they attract different audiences and create different relationships. The encouraging voice builds community through support, while the provocative voice creates engagement through debate and strong opinions.

Your brand voice should align with your natural communication style while being intentionally crafted for your audience's preferences. If your audience appreciates detailed explanations, incorporate thoroughness into your voice. If they prefer quick, actionable tips, develop a more concise and direct communication style.

Document your brand voice with specific examples and guidelines. Create a reference that describes how you handle different types of content, educational posts, behind-the-scenes content, controversial topics, and promotional messages. This documentation ensures consistency across all your content and helps if you ever work with team members or AI tools.

Building Unshakeable Authority

Authority is earned trust demonstrated through consistent value delivery and expertise. It's not about having the most followers or the longest resume. It's about being the person your audience turns to when they need reliable information or solutions in your area of expertise.

Authority builds through three mechanisms: demonstrated knowledge, consistent results, and social proof. Demonstrated knowledge means regularly sharing insights, strategies, and information that help your audience achieve their goals. Consistent results mean practicing what you preach and showing the outcomes of your methods. Social proof means others validating your expertise through testimonials, collaborations, and recognition.

The fastest way to build authority is by solving real problems for real people publicly. Answer questions thoroughly, share case studies from your experience, and provide actionable advice that people can implement immediately. Each piece of helpful content adds to your authority bank account.

CRAFTING YOUR UNIQUE VALUE PROPOSITION (UVP)

Your Unique Value Proposition is the specific benefit you provide that your audience can't get anywhere else, communicated in a way that immediately resonates with their needs and desires. It's the answer to the question, "Why should someone follow you instead of everyone else in your space?"

Most creators struggle with their UVP because they focus on features rather than benefits, or they try to appeal to everyone instead of being indispensable to someone. A strong UVP is specific, beneficial, and unique. It clearly states what you do, who you do it for, and what makes your approach different or better.

The best UVPs emerge from the intersection of your unique experience, your audience's pressing needs, and market gaps that existing creators haven't filled effectively. This intersection is where you can provide maximum value with minimum competition.

The UVP Discovery Process

Start by cataloging your unique combination of experiences, skills, and perspectives. Include professional experience, personal challenges you've overcome, hobbies that inform your expertise, and unusual combinations of knowledge areas. Most people are combinations of several things rather than experts in just one area.

Next, research your audience's specific pain points, goals, and frustrations. Look at comments on other creators' content, questions in relevant forums and groups, and direct messages you receive. Pay attention to the language people use to describe their challenges, this language should appear in your UVP.

Finally, analyze the competitive landscape to identify gaps or opportunities. Look for approaches that aren't being taken, audiences that aren't being served, or ways of communicating that aren't being used. Your UVP should position you in a space where you can be first or best.

UVP Formula and Examples

An effective UVP follows this structure: "I help [specific audience] achieve [specific outcome] through [unique approach/method] so they can [ultimate benefit]." This formula ensures you address who, what, how, and why in a concise statement.

Example 1: "I help overwhelmed small business owners systematize their operations using simple tech solutions so they can work fewer hours while growing revenue." This UVP is specific about the audience (overwhelmed small business owners), the method (simple tech solutions), and the ultimate benefit (work less, earn more).

Example 2: "I teach introverted professionals how to build influential personal brands through strategic content creation so they can advance their careers without networking events or self-promotion." This UVP addresses a specific personality type with a common challenge and offers a solution that aligns with their preferences.

Your UVP should be flexible enough to encompass various content types while being specific enough that your ideal audience immediately recognizes themselves in it. It becomes the filter for all your content decisions. If a piece of content doesn't support your UVP, it doesn't belong in your content strategy.

Testing and Refining Your UVP

Your UVP isn't set in stone. It should evolve as you better understand your audience and refine your expertise. Test different versions through your content and pay attention to which messages generate the strongest response from your ideal audience.

Create content that explicitly states your UVP in different ways and measure engagement levels. Use your bio, about sections, and introductory content to test various phrasings. The version that generates the most qualified followers and engagement is likely closest to your optimal UVP.

Refine your UVP based on feedback from your audience, changes in your expertise, and shifts in market demand. A UVP that worked when you had 1,000 followers might need adjustment when you have 50,000 followers and deeper expertise.

SETTING MEASURABLE GOALS (GROWTH VS. MONETIZATION)

Success in social media requires balancing two often competing objectives: audience growth and revenue generation. Many creators optimize for one at the expense of the other, leading to either large audiences that don't convert or small audiences that generate income but limit long-term potential.

The key is understanding that growth and monetization operate on different timelines and require different strategies. Growth strategies focus on reach, engagement, and follower acquisition. Monetization strategies focus on conversion, customer lifetime value, and revenue per follower. Both are necessary, but their emphasis should shift based on your current stage and objectives.

Setting measurable goals creates accountability and provides clear indicators of what's working and what needs adjustment. Without specific metrics, it's impossible to know whether your efforts are moving you toward your objectives or simply keeping you busy.

The Growth vs. Monetization Balance

In the early stages of building your social media presence, growth should be your primary focus. You need sufficient audience size to make monetization viable, and growth activities help you understand your audience's preferences and refine your content strategy. During this phase, monetization should be minimal and focused on building trust rather than maximizing revenue.

As your audience grows and engagement stabilizes, you can shift toward balanced growth and monetization. This means maintaining growth activities while introducing revenue-generating content and offers. The goal is to monetize your existing audience while continuing to attract new followers.

At mature stages, monetization can become the primary focus, with growth being a byproduct of successful revenue generation. Happy customers become advocates who attract new followers organically. At this stage, you optimize for revenue per follower rather than total follower count.

SMART Goals for Social Media Success

Effective social media goals follow the SMART framework: Specific, Measurable, Achievable, Relevant, and Time-bound. Vague goals like "grow my following" provide no guidance for strategy or tactics. Specific goals like "gain 5,000 Instagram followers in 90 days through daily Reels and strategic hashtag use" provide clear direction and measurable outcomes.

Growth goals should include metrics for followers, engagement rate, reach, and content performance. Track leading indicators (content creation consistency, engagement rate) and lagging indicators (follower growth, website traffic) to understand cause-and-effect relationships in your strategy.

Monetization goals should include revenue targets, conversion rates, average order value, and customer lifetime value. These metrics help you understand not just how much money you're making, but the sustainability and scalability of your revenue streams.

Goal Setting Framework by Stage

Startup Stage (0-1,000 followers):

- Growth Focus: Establish consistent posting schedule, achieve 5% engagement rate, test content formats
- Monetization Focus: Build email list, create lead magnets, establish credibility

Growth Stage (1,000-10,000 followers):

- Growth Focus: Increase posting frequency, achieve viral content, expand to new platforms
- Monetization Focus: Launch first paid offering, establish affiliate partnerships, build customer testimonials

Scale Stage (10,000+ followers):

- Growth Focus: Optimize high-performing content, collaborate with other creators, maintain engagement rates
- Monetization Focus: Develop multiple revenue streams, increase average order value, build recurring income

Each stage requires different strategies and metrics, but the goals should be interconnected. Growth activities should support future monetization, and monetization activities should enhance long-term growth potential.

Tracking and Adjustment Systems

Create a simple tracking system that monitors your key metrics weekly and monthly. Use platform analytics, Google Analytics, and revenue tracking to understand performance trends. Look for patterns that connect content types, posting times, and topics to both growth and monetization outcomes.

Schedule monthly goal review sessions where you analyze performance, identify what's working, and adjust strategies that aren't meeting expectations. Social media moves quickly, so quarterly adjustments might be too infrequent to capitalize on opportunities or address problems.

Set milestone celebrations and course corrections based on your tracking data. If you're ahead of schedule, consider increasing your goals or expanding your strategies. If you're behind schedule; analyze whether the goals were unrealistic or if the strategies need adjustment.

Action Steps for Chapter 3:

Action Item	Task Description	Expected Outcome
Brand Identity Audit	Complete a comprehensive assessment of your unique experiences, values, and expertise. Create a "brand DNA" document that captures what makes you different from competitors in your space.	Clear foundation for all future branding and content decisions
Voice Development	Write in your natural style for 30 minutes daily for one week, then analyze patterns in tone, vocabulary, and structure. Create brand voice guidelines with specific examples and phrases that represent your communication style.	Consistent, recognizable brand voice across all content
UVP Testing	Create three different versions of your Unique Value Proposition and test them in your bio, content introductions, and initial audience interactions. Track which version generates the most engagement from your ideal audience.	Validated UVP that resonates with target audience
Goal Setting Workshop	Define specific, measurable goals for both growth and monetization over the next 90 days. Include leading indicators (content creation, engagement) and lagging indicators (followers, revenue).	Clear roadmap with trackable milestones for next quarter

Action Item	Task Description	Expected Outcome
Tracking System Setup	Establish a simple weekly tracking system for key metrics using spreadsheets or tools like Google Analytics. Schedule weekly reviews and monthly strategy adjustments based on performance data.	Data-driven approach to measuring and improving performance

FOUR

Mastering SEO for Social Media

Search Engine Optimization for social media operates differently than traditional SEO, but the fundamental principle remains the same: understanding what your audience is searching for and creating content that matches their search intent. Each platform has its own search ecosystem with unique ranking factors, user behaviors, and optimization opportunities.

The misconception that social media content doesn't need SEO has cost creators millions of views and potential customers. Every major social platform now functions as a search engine. TikTok's search volume rivals Google for certain demographics, Instagram users search for specific content types and creators, YouTube is the second-largest search engine globally, and Facebook's search function helps users discover relevant groups and content.

Understanding platform-specific keyword strategies allows your content to be discovered weeks, months, or even years after publication. While algorithm-driven feeds provide immediate visibility, search-driven discovery creates lasting value that compounds over time.

TikTok Keyword Strategy

TikTok's search algorithm prioritizes recency, engagement velocity, and keyword relevance in video descriptions, captions, and audio. The platform's younger demographic searches differently than traditional search engine users—they use conversational phrases, trending slang, and emotion-driven keywords.

Instead of searching for "weight loss tips," TikTok users might search for "how to lose belly fat fast" or "workouts that actually work." They often include emotional modifiers like "easy," "quick," "that actually work," or "no one talks about." Incorporating these natural speech patterns into your captions increases discoverability.

TikTok keyword research involves monitoring trending hashtags, analyzing successful content in your niche, and using TikTok's search suggestions feature. Type your main keywords into TikTok's search bar and note the autocomplete suggestions—these represent high-volume search terms that users frequently search for.

The platform also responds to keywords mentioned in your video audio. Speaking your target keywords naturally during your content can improve search visibility, even if those keywords aren't in your written caption. This creates opportunities for organic keyword integration without compromising content quality.

Instagram Search Optimization

Instagram search encompasses hashtags, locations, accounts, and audio searches. The platform's algorithm considers caption keywords, hashtag relevance, location tags, and user interaction patterns when determining search visibility.

Instagram users search for inspiration, solutions, and specific content types. They might search for "small business marketing tips," "sustainable fashion brands," or "healthy meal prep ideas." Your keyword strategy should reflect these intent-driven searches rather than just trending topics.

Effective Instagram keyword integration requires strategic placement in multiple areas: your bio, post captions, hashtags, alt text, and story highlights. Each area carries different weight in the search algorithm, so diversifying your keyword presence increases overall discoverability.

The platform's hashtag system functions as a keyword categorization tool. Research hashtags with varying competition levels, high-volume hashtags for maximum reach, medium-volume hashtags for qualified traffic, and low-volume niche hashtags for targeted audience segments.

YouTube SEO Mastery

YouTube SEO most closely resembles traditional search engine optimization because the platform prioritizes search intent satisfaction and user retention metrics. Keywords in titles, descriptions, tags, and transcripts all influence ranking, but user behavior signals like watch time and engagement carry significant weight.

YouTube keyword research requires understanding both search volume and competition levels. Tools like Google Keyword Planner, TubeBuddy, and VidIQ provide data on search volumes and competition for specific terms. Focus on keywords with sufficient search volume but manageable competition levels for your channel size.

Long-tail keywords often provide better opportunities for smaller channels. Instead of competing for "fitness tips" (high competition), target "fitness tips for busy moms over 40" (lower competition, higher intent). These specific phrases attract viewers who are more likely to engage and convert.

YouTube's search algorithm also considers semantic relationships between keywords. If your main keyword is "productivity," the algorithm recognizes related terms like "time management," "efficiency," "organization," and "workflow optimization." Including semantically related keywords in your content improves topical authority and ranking potential.

Facebook Discovery Optimization

Facebook's search functionality focuses on connecting users with relevant groups, pages, and content within their network. The platform prioritizes content from sources that users have previously engaged with, making relationship-building crucial for long-term search visibility.

Facebook keyword optimization occurs primarily in page descriptions, post text, and group names and descriptions. The platform's algorithm considers engagement patterns, post performance history, and user relationship strength when determining search ranking.

Unlike other platforms, Facebook search often reflects personal networking intent. Users search for industry experts, local businesses, and community groups. Your keyword strategy should include professional terms, location-based keywords, and community-focused language that reflects how your audience thinks about connecting with others in your space.

OPTIMIZING CAPTIONS, DESCRIPTIONS, AND HASHTAGS

The technical elements of your posts, captions, descriptions, and hashtags, function as the infrastructure that determines whether your content gets discovered by your ideal audience. These elements work together to signal to platform algorithms what your content is about, who should see it, and how it should be categorized and ranked.

Optimization doesn't mean keyword stuffing or sacrificing readability for search performance. The best-optimized content seamlessly integrates keywords and technical elements while maintaining natural language flow and providing genuine value to readers. This balance requires understanding each platform's specific requirements and user expectations.

Effective optimization creates multiple pathways for content discovery. A well-optimized post can be found through hashtag searches, keyword searches, algorithmic recommendations, and user shares. Each pathway reaches different segments of your potential audience and serves different discovery intents.

Caption Optimization Strategies

Captions serve dual purposes: engaging your current audience and attracting new viewers through search. The opening lines are crucial because they appear in feed previews and influence whether users engage with your full post. Lead with your most important keywords while maintaining a natural conversational tone.

Structure your captions with search optimization in mind. Place primary keywords in the first sentence, include secondary keywords naturally throughout the text, and end with relevant hashtags and calls-to-action. This structure satisfies both algorithmic requirements and user reading patterns.

Different platforms reward different caption approaches. Instagram favors longer, story-driven captions that encourage extended engagement. TikTok prefers shorter, punchy captions that complement video content. LinkedIn responds well to professional, insight-driven captions that spark discussion.

Consider search intent when crafting captions. Users searching for "budget meal prep ideas" expect practical, actionable content. Users searching for "meal prep inspiration" might prefer aspirational, visually focused content. Your caption tone and structure should match the intent behind your target keywords.

Description Optimization for Video Platforms

YouTube descriptions provide the most comprehensive optimization opportunities among social platforms. The first 125 characters appear in search results and suggested video previews, making them crucial for click-through rates. Use this space to clearly communicate your video's value proposition while incorporating primary keywords.

Structure YouTube descriptions with multiple sections: hook (first 125 characters), detailed explanation of video content, timestamps for easy navigation, links to relevant resources, and additional context or background information. This structure serves both search algorithms and user experience needs.

TikTok descriptions are limited but powerful for search optimization. Include your primary keyword early in the description, add 2-3 relevant hashtags, and include a clear call to action. The brevity forces you to be precise with keyword selection and messaging.

Video descriptions should also consider related content optimization. Include keywords for content series, recurring themes, or topic clusters you're building. This helps establish topical authority and create connections between your different pieces of content.

Strategic Hashtag Implementation

Hashtags function as content categorization systems that help platforms understand and distribute your content to relevant audiences. Each platform treats hashtags differently, requiring tailored strategies for optimal performance.

Instagram hashtag strategy should include a mix of high-volume, medium-volume, and niche hashtags. Use 5-10 hashtags per post, focusing on relevance over popularity. Research hashtags by examining their recent posts; active, engaged communities indicate valuable hashtag opportunities.

TikTok hashtags work differently, functioning more like keywords than content categories. Use three to five highly relevant hashtags that describe your content and target audience. Steer clear of excessively popular hashtags that could potentially obscure your content amidst millions of posts.

LinkedIn hashtags should be professional and industry-specific. Use 3-5 hashtags maximum, focusing on business terms, industry keywords, and professional development topics. LinkedIn users search for career-relevant content, so your hashtags should reflect professional value.

Create branded hashtags for community building and content organization. Branded hashtags help your audience find related content and encourage user-generated content. Choose memorable, relevant phrases that represent your brand values or content themes.

Advanced Optimization Techniques

Alt text optimization for images provides additional keyword opportunities while improving accessibility. Describe your images naturally while incorporating relevant keywords. This text isn't visible to most users but helps search algorithms understand your content's context.

Geolocation tags expand your content's discoverability to location-based searches. Even if your content isn't location-specific, relevant location tags can increase visibility among local audiences who might become customers or collaborators.

Cross-platform optimization involves adapting your content for each platform's specific requirements while maintaining consistent keyword themes. A single piece of content can be optimized differently for Instagram, TikTok, YouTube, and LinkedIn while targeting the same core keywords.

Monitor hashtag and keyword performance regularly. Platforms provide analytics on how users discover your content, which hashtags drive the most impressions, and which keywords generate the most engagement. Use this data to refine your optimization strategies continuously.

HOW SEO DRIVES DISCOVERABILITY & LONG-TERM REACH

SEO-optimized social media content creates compound growth that extends far beyond initial publication. While algorithm-driven distribution provides immediate visibility, search-optimized content continues attracting new audiences for months or years after creation. This long-term discoverability transforms content from consumable entertainment into valuable business assets.

Understanding the relationship between immediate performance and long-term search value helps creators balance viral content strategies with sustainable growth approaches. Viral content provides short-term visibility boosts, but SEO-optimized content delivers consistent, qualified traffic that often converts better than algorithm-driven views.

The compounding effect of search-optimized content means that each piece of content adds to your overall discoverability. As you build a library of search-friendly content, your total search presence grows exponentially rather than linearly. This creates a competitive advantage that becomes stronger over time.

The Long-tail Discovery Effect

Most social media creators focus on trending topics and viral content opportunities, leaving significant long-tail search opportunities underserved. Long-tail keywords, specific, detailed search phrases, often have lower competition and higher conversion intent than broad, popular terms.

For example, instead of creating content around "marketing tips" (high competition), you might target "email marketing tips for small service businesses" (lower competition, higher intent). These specific searches attract users with clear needs who are more likely to engage meaningfully with your content and become customers.

Long-tail content discovery happens through various pathways: direct keyword searches, related content recommendations, and cross-platform search results. Google often displays social media content in search results, meaning your optimized posts can attract traffic from traditional search engines in addition to platform-specific searches.

Building a content library around long-tail keywords creates multiple entry points for new audience members. Each piece of content targets specific search intents, and collectively they establish you as an authority on broader topic areas. This approach builds sustainable growth rather than dependence on viral moments.

Search-Driven Community Building

SEO-optimized content attracts audiences based on shared interests and needs rather than just algorithmic matching. This creates higher-quality community connections because people find your content while actively seeking solutions you provide.

Search-discovered audiences tend to be more engaged because they've demonstrated intent by searching for specific information. They're not passive scroll victims who happened to see your content. They're active searchers who found exactly what they were looking for in your posts.

This quality of audience discovery leads to better conversion rates for your monetization efforts. Followers who discovered you through search are more likely to purchase products, join email lists, and recommend your content to others with similar needs.

Search-optimized content also supports community building through content connections. When users search for related topics, your interconnected content appears as a comprehensive resource rather than isolated posts. This positions you as a go-to expert rather than just another content creator.

Evergreen Content Strategy

Evergreen content, material that remains relevant over time, benefits most from SEO optimization because it continues attracting search traffic long after publication. This type of content includes tutorials, explanations of fundamental concepts, and solutions to persistent problems.

Developing an evergreen content strategy requires identifying topics in your niche that maintain search volume over time. While trending topics provide temporary traffic spikes, evergreen topics provide consistent, predictable traffic that supports long-term business growth.

Evergreen SEO content should be comprehensive, accurate, and regularly updated to maintain search relevance. Platforms favor content that continues generating engagement over time, so maintaining and refreshing your evergreen content improves its long-term search performance.

The ROI of evergreen content increases over time as search rankings improve and content authority builds. A single piece of well-optimized evergreen content can generate more total value than dozens of trending topic posts that quickly lose relevance.

Cross-Platform SEO Synergy

Your SEO efforts across different platforms should work together to build comprehensive topic authority. When you consistently create optimized content around specific themes across multiple platforms, you establish expertise signals that benefit all your content.

Cross-platform synergy also creates multiple pathways for audience discovery. A user might find your YouTube video through search, discover your Instagram account through your video description, and join your email list through your Instagram bio. Each platform serves as a discovery point that leads to your broader content ecosystem.

Repurposing content across platforms with platform-specific optimization maximizes the SEO value of each content idea. A single concept can become a YouTube video, an Instagram carousel, a TikTok series, and a LinkedIn article, each optimized for platform-specific search behaviors while reinforcing your expertise on the topic.

Track cross-platform performance to understand how SEO efforts on one platform influence growth on others. Users who discover you through search on one platform often become followers across multiple platforms, creating compound growth effects that amplify your total reach and influence.

Action Steps for Chapter 4:

Action Item	Task Description	Expected Outcome
Platform Keyword Research	Use each platform's search function to identify 20 high-potential keywords in your niche. Create a keyword database with search volume estimates and competition levels for each platform.	Comprehensive keyword strategy tailored to each platform's search behavior
Content Audit & Optimization	Review your last 20 posts across all platforms and optimize captions, descriptions, and hashtags using your keyword research. Focus on incorporating keywords naturally while maintaining readability.	Immediate improvement in content discoverability and search performance
Hashtag Strategy Development	Create platform-specific hashtag banks with 50 relevant hashtags categorized by competition level (high, medium, low volume). Include branded hashtags and location tags where appropriate.	Strategic hashtag system for consistent content optimization
Evergreen Content Calendar	Plan 10 pieces of evergreen content around your top keywords. Create titles, outlines, and optimization strategies for content that will remain relevant for 6+ months.	Foundation for long-term search traffic and sustainable growth
SEO Performance Tracking	Set up tracking systems to monitor how users discover your content (search vs. algorithm vs. direct). Use platform analytics to identify which keywords and optimization strategies drive the best results.	Data-driven insights to continuously improve SEO effectiveness

FIVE

TikTok Domination

TikTok's algorithm is both the most democratic and most unpredictable among social media platforms. Unlike other platforms that heavily favor established creators with large followings, TikTok's algorithm gives every video a chance to go viral, regardless of the creator's follower count. This creates unprecedented opportunities for new creators to build massive audiences quickly, but it also requires understanding the specific factors that influence content distribution.

The algorithm operates on a recommendation system that evaluates each video within the first few hours of publication. During this critical window, the platform tests your content with small audience segments and measures engagement velocity, how quickly users like, comment, share, and watch your video completely. High engagement signals prompt the algorithm to show your content to progressively larger audiences.

Understanding these mechanics allows you to optimize your content for algorithmic success rather than leaving performance to chance. Every element of your video, from the first three seconds to the final call-to-action, should be designed to maximize the engagement signals that drive algorithmic distribution.

The Four Pillars of TikTok's Algorithm

User interaction forms the foundation of TikTok's recommendation system. The algorithm tracks which accounts users follow, which videos they like and share, which creators they engage with through comments, and which content they watch to completion. Your content performs better when it generates the types of interactions that TikTok values most.

Video information includes details like captions, hashtags, music, and visual effects. The algorithm uses this metadata to understand your content's topic and match it with users interested in similar content. Strategic use of trending sounds, relevant hashtags, and descriptive captions helps the algorithm categorize and distribute your content effectively.

Device and account settings influence content distribution based on language preferences, location, and device type. While you can't control these factors for your audience, understanding them helps you create content that appeals to your target demographic within TikTok's geographic and linguistic parameters.

Content completion rates and engagement velocity determine how widely the algorithm distributes your content beyond your immediate followers. Videos that maintain viewer attention throughout their duration and generate immediate engagement receive broader distribution than content that loses viewers quickly or generates delayed responses.

Cracking the Engagement Velocity Code

The first hour after posting is crucial for TikTok success. During this period, the algorithm evaluates how your content performs with your existing followers and small test audiences. Strong performance signals, high completion rates, immediate likes and comments, quick shares, trigger broader distribution to users who don't follow you.

Optimization for engagement velocity requires strategic timing and immediate promotion. Post when your audience is most active, and promote your new content through other platforms, direct messages, and community engagement immediately after publishing. The goal is to generate as much engagement as possible within the first 60 minutes.

Creating "binge-worthy" content that encourages users to watch multiple times or visit your profile immediately after viewing can boost velocity metrics. Content that sparks curiosity, provides incomplete information that requires profile visits, or creates emotional responses that prompt immediate engagement performs better in the velocity evaluation.

Monitor your content's performance closely during the first few hours. If a video isn't generating expected engagement within the first hour, promote it more aggressively through comments on other creators' content, shares to other platforms, or direct outreach to your most engaged followers.

The Psychology of TikTok Consumption

TikTok users consume content differently than users on other platforms. They expect immediate entertainment or value, have high attention spans, and scroll quickly through content that doesn't immediately capture their interest. This consumption pattern requires content that delivers value within the first few seconds and maintains engagement throughout.

The platform's infinite scroll design creates addictive consumption patterns where users watch hundreds of videos in single sessions. Your content competes not just with other creators in your niche, but with every piece of entertaining or engaging content on the platform. This reality demands that your content matches or exceeds the entertainment value of viral content while delivering your educational or promotional message.

Understanding TikTok's unique culture and language is crucial for success. The platform has developed its own vocabulary, humor styles, content formats, and community norms that differ significantly from other social media platforms. Content that feels native to TikTok performs better than content that appears to be repurposed from other platforms.

TikTok users also have different expectations for creator authenticity. They prefer raw, unpolished content over highly produced material, and they respond better to creators who participate in trends and community conversations rather than broadcasting one-way promotional messages.

SHORT-FORM CONTENT THAT CONVERTS

Creating TikTok content that drives business results requires balancing entertainment value with clear conversion objectives. The most successful business-focused TikTok creators understand that direct sales content rarely succeeds on the platform, but entertaining or educational content that subtly guides viewers toward business objectives can generate significant results.

The key to conversion-focused TikTok content is providing immediate value while creating natural pathways for deeper engagement with your brand. This might mean teaching a valuable skill while mentioning your paid course, sharing a personal transformation story that highlights your coaching services, or creating entertaining content that showcases your expertise.

Conversion on TikTok often happens through indirect pathways rather than direct calls-to-action. Users might watch your content, visit your profile, click your bio link, join your email list, and eventually purchase your products days or weeks later. This extended conversion timeline requires content strategies that build trust and authority over time rather than pushing immediate sales.

The Hook-Value-CTA Framework

Every successful TikTok video follows a predictable structure that captures attention, delivers value, and guides viewers toward desired actions. The hook (first 3-5 seconds) determines whether users continue watching or scroll to the next video. The value section (middle of the video) provides the educational or entertainment content that justifies the viewer's time investment. The call to action (final 5-10 seconds) guides viewers toward deeper engagement with your brand.

Effective hooks create curiosity gaps, make bold statements, ask compelling questions, or promise specific benefits. "The mistake that's costing you thousands in lost revenue" creates more curiosity than "Marketing tips for small businesses." Your hook should be specific enough to attract your ideal audience while being intriguing enough to capture broader interest.

The value section must deliver on your hook's promise while being substantial enough to satisfy viewers. This doesn't mean providing comprehensive tutorials in 60 seconds, but rather giving actionable insights, useful tips, or entertaining content that leaves viewers feeling like they gained something valuable from watching your video.

Call-to-actions on TikTok work best when they feel natural rather than forced. Instead of "Follow me for more tips," try "What other business challenges should I cover?" This approach encourages engagement while subtly requesting the algorithmic boost that comes from comments and follows.

Content Formats That Drive Business Results

Tutorial and educational content consistently performs well for business-focused creators because it provides clear value while demonstrating expertise. Break complex processes into simple steps, use visual demonstrations, and focus on actionable advice that viewers can implement immediately.

Behind-the-scenes content builds trust and authenticity while showcasing your expertise in natural contexts. Share your work process, decision-making approaches, client interactions (with permission), or the reality of running your business. This content type humanizes your brand and creates emotional connections with potential customers.

Transformation and case study content provides social proof while entertaining viewers. Share before-and-after results, client success stories, or your journey of overcoming challenges related to your expertise. These stories demonstrate your effectiveness while inspiring viewers who face similar challenges.

Trend-jacking involves adapting popular TikTok trends, sounds, or formats to your business message. This approach leverages existing viral momentum while ensuring your content feels native to the platform. The key is maintaining the entertainment value of the original trend while incorporating your unique perspective or expertise.

The Soft Conversion Strategy

Direct selling rarely works on TikTok, but soft conversion techniques can generate significant business results without alienating the platform's entertainment-focused audience. Soft conversion means providing value first and allowing viewers to naturally discover your business offerings through subtle integration and profile optimization.

Educational content that solves problems related to your paid services demonstrates your expertise while helping potential customers understand the value you provide. Instead of promoting your marketing course, create content that teaches specific marketing strategies while mentioning that you help businesses implement comprehensive marketing systems.

Personal branding through consistent content themes, visual style, and messaging creates subconscious associations between your personality and your expertise. When viewers consistently see valuable content from you, they begin to perceive you as an authority in your field, making them more likely to seek out your paid offerings.

Profile optimization plays a crucial role in soft conversion. Your bio, link-in-bio page, and profile content should clearly communicate your expertise and offerings without being overly promotional. Use your bio to clearly state what you do and who you help, and ensure your link directs to a page that captures visitor information and presents your offerings effectively.

TIKTOK SHOP & MONETIZATION METHODS

TikTok's built-in monetization features have evolved rapidly, providing creators with multiple pathways to generate revenue directly through the platform. TikTok Shop, Creator Fund, Live Gifts, and brand partnership tools offer opportunities for immediate monetization, while the platform's massive reach creates opportunities for indirect revenue through audience building and traffic direction.

Understanding TikTok's monetization ecosystem allows creators to develop comprehensive revenue strategies that combine platform-native monetization with external business development. The most successful creators use TikTok's built-in features to generate immediate income while building audiences that support larger business objectives.

The platform's monetization features continue expanding rapidly, with new tools and opportunities launching regularly. Staying current with these developments and testing new features early can provide significant competitive advantages as TikTok continues prioritizing creator economy growth.

TikTok Shop Integration Strategies

TikTok Shop allows creators to sell products directly through the platform, with integrated checkout, inventory management, and customer service features. The platform promotes Shop-enabled content more aggressively because it generates direct revenue for TikTok through transaction fees.

Successful TikTok Shop strategies focus on products that demonstrate well in short-form video content. Physical products with visible before-and-after results, items that solve specific problems, and products that create satisfying visual content tend to perform better than complex or abstract offerings.

Product integration should feel natural rather than forced. Create content that showcases your products in use, demonstrates their benefits, or tells stories about why you created them. The most successful Shop creators treat their products as characters in their content rather than items to be sold.

Shop performance benefits from consistent product-focused content rather than occasional promotional posts. Dedicate specific content types to product demonstration, customer testimonials, and behind-the-scenes production content. This approach builds product awareness and purchase intent over time rather than relying on single promotional pushes.

Creator Fund and Live Monetization

TikTok's Creator Fund pays creators based on video views and engagement, providing passive income for consistent content creators. While individual payments are typically small, creators with millions of views can generate significant monthly income through the program.

Creator Fund optimization requires understanding the program's payment algorithms and creating content that maximizes both views and engagement. Educational content, trending topic coverage, and consistent posting schedules tend to generate the sustained viewership that drives Fund payments.

Live streaming on TikTok provides opportunities for real-time audience engagement and gift-based monetization. Viewers can purchase and send virtual gifts during live streams, which convert to real money for creators. Successful live streaming requires interactive content, regular streaming schedules, and community building that encourages recurring viewership.

Live monetization strategies include educational sessions, Q&A periods, behind-the-scenes content, and collaborative streams with other creators. The key is providing value that justifies viewers' time investment and gift-giving, rather than simply asking for support.

External Monetization Through TikTok Traffic

While TikTok's native monetization features provide direct revenue opportunities, the platform's greatest value for many creators lies in audience building and traffic direction to external monetization channels. Email list building, website traffic generation, and brand awareness development can drive significantly more revenue than platform-native features.

Link-in-bio optimization is crucial for external monetization success. Your bio link should direct to a landing page that captures visitor information and presents your offerings clearly. Use tools like Linktree, Beacons, or custom landing pages to create conversion-optimized destinations for your TikTok traffic.

Email list building through TikTok requires lead magnets that appeal to your audience while being deliverable through digital formats. Free guides, templates, checklists, or exclusive content can incentivize email signups from engaged TikTok viewers who want deeper access to your expertise.

Cross-platform promotion uses TikTok to drive traffic to other social media platforms, websites, or online businesses where monetization may be easier or more profitable. This strategy treats TikTok as a top-of-funnel marketing channel rather than a direct revenue source.

Brand Partnership and Sponsored Content

Brand partnerships represent one of the most lucrative monetization opportunities for TikTok creators with engaged audiences. Brands increasingly recognize TikTok's influence on purchasing decisions and are willing to pay significant fees for authentic creator endorsements and product placements.

Partnership opportunities range from one-off sponsored posts to long-term brand ambassadorships. Rates vary significantly based on audience size, engagement rates, niche relevance, and content quality. Creators with 100,000 engaged followers can often command $1,000-5,000 per sponsored post, while top creators earn much more.

Building brand partnership opportunities requires demonstrating audience engagement, maintaining content quality, and developing a professional approach to creator business relationships. Create a media kit that showcases your audience demographics, engagement statistics, and previous partnership successes.

Authentic integration is crucial for brand partnership success on TikTok. Audiences can easily detect overly promotional content and may respond negatively to creators who sacrifice authenticity for sponsorship revenue. The most successful brand partnerships feel like natural extensions of the creator's regular content rather than obvious advertisements.

Action Steps for Chapter 5:

Action Item	Task Description	Expected Outcome
Algorithm Optimization Audit	Analyze your last 10 TikTok posts for engagement velocity patterns. Identify posting times, content types, and hooks that generate the fastest initial engagement. Create a posting strategy based on your highest-performing patterns.	Data-driven posting strategy optimized for TikTok's algorithm preferences
Content Framework Development	Create 20 video ideas using the Hook-Value-CTA framework. Write specific hooks, outline value delivery methods, and design natural call-to-actions for each concept. Focus on soft conversion approaches.	Systematic approach to creating conversion-focused content that feels native to TikTok
TikTok Shop Setup	If applicable to your business, set up TikTok Shop and create product-focused content for your top 3 offerings. If not applicable, optimize your bio link and create lead magnets for email list building.	Direct monetization pathway through TikTok's native commerce features
Trend Integration Strategy	Identify 10 current trends in your niche and adapt them to showcase your expertise. Create content that participates in trends while maintaining your educational or business focus.	Increased visibility through trend participation while maintaining brand integrity
Performance Tracking System	Establish metrics tracking for TikTok-specific goals: engagement velocity, completion rates, profile visits, and conversion from TikTok to external platforms. Monitor these metrics weekly to optimize performance.	Clear understanding of what drives both TikTok success and business results

SIX

Instagram Influence

Instagram's multi-format ecosystem provides creators with diverse ways to connect with audiences, but each format serves different purposes and requires distinct strategies. Reels drive discovery and viral growth, Stories build intimate daily connections, and Live streams create real-time community engagement. Mastering all three formats creates a comprehensive growth strategy that captures audiences at different engagement levels and consumption preferences.

The platform's algorithm prioritizes creators who utilize multiple content formats consistently, viewing this as a signal of serious commitment to the platform. Creators who post only feed content or only Reels miss opportunities for algorithmic favor and audience connection. The most successful Instagram influencers treat each format as part of an integrated content ecosystem where different formats serve specific functions in their overall growth strategy.

Understanding how each format contributes to your growth objectives allows you to allocate time and energy efficiently. Reels build new audience awareness, Stories nurture existing relationships, and Lives convert engaged followers into customers or deeper community members. This strategic approach prevents content creation overwhelm while maximizing results from each piece of content you create.

Reels Growth Acceleration

Instagram Reels represent the platform's biggest growth opportunity because they receive preferential algorithmic distribution and appear on the dedicated Reels tab, Explore page, and user feeds. The format's short-form video structure rewards engaging, entertaining content that captures attention immediately and maintains interest throughout the video duration.

Successful Reels follow predictable patterns that maximize engagement and shareability. Strong openings that create curiosity or promise value, clear visual storytelling that works without sound, trending audio that leverages existing viral momentum, and compelling captions that encourage comments and shares. Each element works together to signal engagement value to Instagram's algorithm.

Trending audio plays a crucial role in Reels discovery because Instagram promotes content that uses popular sounds to capitalize on existing interest patterns. However, the most successful creators adapt trending audio to their niche rather than copying viral content exactly. This approach captures trend momentum while maintaining brand relevance and authenticity.

Reels optimization also involves strategic timing and technical quality. Post when your audience is most active, ensure good lighting and clear audio, use text overlays to reinforce your message, and include strong calls-to-action that encourage profile visits. These technical elements may seem minor, but they significantly impact performance in Instagram's competitive content environment.

Key Reels Growth Tactics:

- **Trend Adaptation**: Monitor trending audio and effects, then adapt them to your niche with unique perspectives or expertise integration
- **Hook Optimization**: Test different opening lines and visual hooks to identify what captures attention most effectively in your specific niche
- **Series Content**: Create recurring Reels series (like "Monday Motivation" or "Transformation Tuesday") to build audience anticipation and consistency
- **Cross-Promotion**: Share your best-performing Reels to your Stories and feed to maximize visibility and engagement across all formats
- **Community Audio**: Use trending sounds from your niche community rather than just mainstream trends for more targeted audience growth
- **Educational Entertainment**: Balance informational content with entertainment value to satisfy both algorithm preferences and audience retention

Stories for Daily Connection

Instagram Stories provide opportunities for authentic, unpolished content that builds stronger relationships with existing followers. The format's temporary nature encourages more casual, personal sharing that humanizes your brand and creates emotional connections that feed posts and Reels can't replicate.

Stories also offer unique engagement features like polls, questions, quizzes, and shopping stickers that drive direct interaction with your audience. These features provide valuable audience insights while boosting your content's engagement signals, which can positively impact your overall account's algorithmic performance.

The key to successful Stories strategy is consistency and variety. Post Stories daily to maintain top-of-mind awareness with your followers, and use different content types to keep your Stories engaging. Mix behind-the-scenes content, educational tips, personal updates, user-generated content, and promotional messages to create diverse value for your audience.

Stories analytics provide detailed insights into audience behavior, showing you which content types generate the most engagement, profile visits, and link clicks. Use this data to refine your Stories strategy and understand what resonates most with your specific audience.

Live Streaming for Community Building

Instagram Live provides real-time interaction opportunities that create deeper connections with your most engaged followers. The format's immediacy and interactivity make it powerful for building community, answering questions, demonstrating expertise, and converting followers into customers or email subscribers.

Live streaming success requires planning and consistency. Schedule regular Live sessions so your audience knows when to expect them, prepare talking points or activities to avoid awkward silences, and promote upcoming Lives through your other content formats to ensure good attendance.

The intimate nature of Live streaming makes it ideal for high-value activities like coaching calls, Q&A sessions, product demonstrations, or behind-the-scenes business insights. These applications provide immediate value to attendees while positioning you as an accessible expert in your field.

Live content also receives algorithmic priority during broadcast and can be saved as IGTV content for ongoing discovery. This dual benefit means Live streaming provides both immediate engagement and long-term content value.

BUILDING RELATIONSHIPS THROUGH DMS AND COMMUNITIES

Direct Messages and community engagement represent Instagram's most underutilized growth strategies. While most creators focus exclusively on content creation, the most successful influencers understand that Instagram is fundamentally a relationship-building platform where personal connections drive long-term success more than viral content.

DM strategies and community engagement require more time investment than broadcast content creation, but they generate higher conversion rates and stronger audience loyalty. Followers who feel personally connected to creators are more likely to purchase products, recommend accounts to friends, and engage consistently with content over time.

Building genuine relationships at scale requires systematic approaches that maintain authenticity while managing time efficiently. This means developing templates for common responses, creating engagement schedules, and using Instagram's features strategically to nurture relationships without overwhelming your daily schedule.

Strategic DM Engagement

Direct message engagement should focus on your most valuable followers—those who consistently engage with your content, share your posts, or express interest in your offerings. These power users often become brand advocates, customers, or collaborative partners when nurtured through personal connection.

Effective DM strategies include thanking new followers who engage meaningfully with your content, responding thoughtfully to comments and questions, reaching out to users who share your content, and maintaining conversations with potential collaborators or customers. Each interaction should feel personal rather than automated.

Use DMs to move relationships from public comments to private conversations where deeper connection and conversion opportunities exist. When someone asks detailed questions in comments, invite them to DM for more comprehensive help. This approach provides better service while creating opportunities for relationship building.

DM automation tools can help manage volume while maintaining personal touch, but avoid overly automated responses that feel impersonal. Use automation for initial outreach or standard information sharing, but switch to personal responses for meaningful conversations.

Community Building Through Engagement

Instagram community building extends beyond your own content to include meaningful participation in your niche's broader conversation. Engage authentically with other creators' content, participate in relevant hashtag communities, and contribute value to discussions in your field.

Strategic community engagement involves identifying key accounts in your niche—both competitors and complementary creators—and building relationships through consistent, valuable interaction. Comment meaningfully on their posts, share their content when relevant, and look for collaboration opportunities that benefit both parties.

User-generated content campaigns can transform followers into active community participants who create content featuring your brand, products, or expertise. These campaigns provide social proof while reducing your content creation burden and strengthening community connections.

Host community events like hashtag challenges, Q&A sessions, or collaborative Lives with other creators to bring your audience together around shared interests. These events create memorable experiences that strengthen community bonds and increase account loyalty.

Conversion Through Relationship Building

The strongest Instagram conversions come from relationships rather than sales pitches. Followers who trust you personally are more likely to purchase your products, join your email list, or recommend your services to others. This relationship-first approach requires patience but generates higher lifetime customer value.

Build trust through consistent value delivery, authentic personal sharing, and reliable communication. When followers see you as a helpful expert rather than just another marketer, they become more receptive to your offers and recommendations.

Use Stories and DMs to nurture potential customers through longer sales cycles. Share case studies, client testimonials, and behind-the-scenes content that demonstrates your expertise and results. This soft-sell approach builds desire without creating sales pressure.

Create exclusive opportunities for your most engaged followers—early access to products, special pricing, or bonus content. These gestures reward loyalty while creating incentives for deeper engagement and community participation.

TURNING FOLLOWERS INTO PAYING CLIENTS/CUSTOMERS

The ultimate goal of Instagram influence is converting social media followers into paying customers for your products, services, or affiliate recommendations. This conversion process requires strategic nurturing that builds trust, demonstrates value, and creates natural pathways from social media engagement to business transactions.

Most creators struggle with conversion because they treat Instagram as a broadcasting platform rather than a relationship-building and sales funnel. Effective conversion strategies integrate social proof, value demonstration, and clear purchasing pathways while maintaining the authentic connections that attract followers in the first place.

Understanding the Instagram-to-customer journey helps optimize each stage of the conversion process. Awareness comes through content discovery, interest develops through consistent value delivery, consideration builds through trust and social proof, and conversion happens through clear offers and simplified purchasing processes.

The Value-First Sales Funnel

Successful Instagram conversion starts with providing significant value before asking for any purchase. This value-first approach builds trust and demonstrates expertise while qualifying potential customers who are genuinely interested in what you offer.

Create content that solves real problems for your target customers, shares case studies of successful outcomes, and provides actionable insights they can implement immediately. This approach positions you as a valuable resource rather than just another person selling something on social media.

Use Instagram's features strategically to guide followers through your sales funnel. Stories can share behind-the-scenes content that builds trust, Reels can demonstrate your expertise, feed posts can provide detailed value, and IGTV can offer comprehensive tutorials that establish authority.

Your bio and link-in-bio page serve as crucial conversion points where interested followers can learn more about your offerings. Optimize these elements to clearly communicate what you do, who you help, and how followers can work with you or purchase your products.

Social Proof and Authority Building

Instagram provides multiple opportunities to showcase social proof that builds confidence in potential customers. Client testimonials, case study posts, before-and-after transformations, and user-generated content all demonstrate that others have succeeded with your help.

Share specific results and outcomes rather than vague testimonials. Numbers, percentages, and concrete achievements are more convincing than general statements about satisfaction or happiness. "Helped Sarah increase her revenue by 150% in 6 months" is more powerful than "Sarah loves working with me."

Feature client successes regularly through dedicated posts, Stories highlights, and video testimonials. This consistent social proof reinforcement helps overcome purchase objections and builds confidence in your abilities to deliver results.

Collaborate with other respected figures in your industry to build authority through association. Guest appearances, joint Lives, and cross-promotion with established experts can boost your credibility with their audiences while expanding your reach.

Clear Conversion Pathways

Remove friction from your Instagram-to-customer journey by creating clear, simple pathways for interested followers to become paying customers. Complex purchasing processes or unclear next steps prevent conversions even when followers are ready to buy.

Use Instagram's shopping features, link stickers, and bio links strategically to guide followers to your offers. Each pathway should be optimized for mobile users and require minimal steps between interest and purchase completion.

Create urgency and scarcity appropriately to encourage timely decision-making. Limited-time offers, exclusive bonuses for Instagram followers, or early-bird pricing can motivate action while rewarding your social media audience.

Follow up with interested prospects who don't convert immediately. Use DMs, email sequences, or remarketing to nurture relationships with followers who have shown purchase interest but need additional time or information to make decisions.

Monetization Strategy Integration

Integrate multiple monetization strategies to maximize revenue from your Instagram influence. Combine product sales, service offerings, affiliate marketing, and sponsored content to create diverse income streams that reduce dependence on any single revenue source.

Price your offerings strategically based on the value you provide and your audience's spending capacity. Instagram audiences often prefer premium products and services when they're positioned appropriately and deliver genuine value.

Create ascending value ladders that allow customers to start with smaller purchases and gradually invest more as they experience success. This approach maximizes customer lifetime value while making your offerings accessible to followers at different financial levels.

Track conversion metrics carefully to understand which content types, offers, and strategies generate the best results. Use Instagram Insights, Google Analytics, and sales tracking to measure ROI and optimize your conversion strategies continuously.

Action Steps for Chapter 6:

Action Item	Task Description	Expected Outcome
Multi-Format Content Strategy	Create a weekly content calendar incorporating daily Stories, 4–5 Reels, and 1–2 Live sessions. Plan content themes that work across all formats while serving different audience engagement purposes.	Systematic approach to utilizing all Instagram formats for maximum algorithmic favor and audience connection

Action Item	Task Description	Expected Outcome
DM Engagement System	Develop templates for common DM responses while maintaining personal touch. Create a daily engagement routine that includes responding to comments, reaching out to power users, and nurturing potential customer relationships.	Scalable relationship-building system that converts followers into customers through personal connection
Community Building Campaign	Launch a user-generated content campaign or recurring hashtag that encourages follower participation. Engage consistently with other creators in your niche to build reciprocal relationships and expand reach.	Active community around your brand that creates content and advocates for your offerings
Conversion Pathway Optimization	Audit your bio, link-in-bio page, and Stories highlights to ensure clear pathways from Instagram to your offers. Create lead magnets specific to Instagram traffic and test different conversion strategies.	Streamlined system for converting Instagram followers into email subscribers and paying customers
Social Proof Content Bank	Collect and organize client testimonials, case studies, and transformation stories into a content bank. Create systematic sharing schedule for social proof content across all Instagram formats.	Consistent social proof reinforcement that builds authority and overcomes purchase objections

SEVEN

Facebook for Business & Passive Income

Facebook's mature ecosystem provides unparalleled opportunities for building sustainable business infrastructure and generating passive income streams. Unlike newer platforms focused on entertainment and trending content, Facebook excels at community building, relationship nurturing, and systematic business development that compounds over time.

The platform's sophisticated advertising system, robust group functionality, and established user behavior patterns make it ideal for creators ready to transition from audience building to serious business development. Facebook users expect more business-oriented content and are generally more willing to engage with promotional material when it provides genuine value.

Understanding Facebook's three primary business tools, Groups for community building, Pages for brand presence, and Ads for targeted reach, allows you to create integrated strategies that work together to build authority, nurture relationships, and generate consistent revenue streams.

Facebook Groups: Community Monetization

Facebook Groups represent the platform's most powerful tool for building engaged communities around your expertise. Unlike Pages, which operate on broadcast models, Groups facilitate genuine discussion, peer-to-peer learning, and community relationships that create strong foundations for business development.

Successful Groups require consistent value delivery through expert insights, facilitated discussions, and member support. The goal isn't to promote your services constantly but to establish yourself as the go-to expert in your niche while building relationships with potential customers over time.

Group monetization happens through various mechanisms: direct service sales to members who need individual help, product recommendations that solve common member problems, affiliate partnerships for tools and resources that benefit the community, and premium membership tiers for enhanced access or additional services.

The key to Group success is maintaining focus on member value rather than personal promotion. Groups that become overly promotional lose engagement and violate Facebook's community guidelines. The most profitable Groups feel like valuable educational communities where the owner happens to offer related services.

Facebook Pages: Professional Presence

Facebook Pages serve as your business headquarters on the platform, providing credibility, contact information, and a professional presence that builds trust with potential customers. Pages also integrate with Facebook's advertising system and provide detailed analytics about your audience and content performance.

Effective Page management involves consistent posting schedules, professional visuals and branding, clear contact information and service descriptions, customer testimonials and reviews management, and regular engagement with comments and messages. These elements work together to create a polished business presence.

Page content should balance educational value with business promotion more directly than Group content. Users who like your Page expect to learn about your services and offerings, making Pages appropriate for showcasing your work, sharing case studies, and promoting your products or services.

Use Pages strategically to support your other marketing activities. Cross-promote your Groups, share your best content from other platforms, highlight client successes, and provide updates about your business developments or new offerings.

Strategic Facebook Advertising

Facebook's advertising platform provides precise targeting capabilities that allow you to reach specific demographics, interests, behaviors, and even custom audiences based on your existing customer data. This precision makes Facebook ads extremely cost-effective when used strategically for business development.

Successful Facebook advertising starts with clear objectives: lead generation for your email list, traffic to your website or blog, engagement with your content, or direct sales of your products or services. Each objective requires different ad formats, targeting strategies, and optimization approaches.

The platform's sophisticated tracking and retargeting capabilities allow you to nurture prospects through extended sales cycles. You can show different ads to people based on their previous interactions with your content, creating personalized advertising sequences that build trust and increase conversion likelihood.

Advanced Facebook Business Strategies:

1. **Lookalike Audience Development**: Upload your best customer email list to Facebook to create lookalike audiences of users with similar characteristics. This targeting approach often generates higher conversion rates than interest-based targeting because it's based on actual customer data rather than assumed preferences.
2. **Retargeting Funnel Creation**: Set up pixel tracking on your website to retarget visitors with specific ads based on pages they visited. Create different ad sequences for people who viewed your services page, downloaded your lead magnet, or abandoned checkout processes to maximize conversion opportunities.
3. **Video View Optimization**: Use video content to build custom audiences of engaged viewers, then retarget these audiences with conversion-focused ads. Video viewers often convert better than cold audiences because they've already invested time in consuming your content and getting to know your expertise.
4. **Cross-Platform Integration**: Use Facebook ads to promote your success on other platforms, driving traffic to your TikTok, Instagram, or YouTube accounts to accelerate growth across your entire social media presence. This strategy leverages Facebook's targeting precision to find audiences likely to engage with your content on other platforms.

FACEBOOK AS A TRUST-BUILDING FUNNEL

Facebook's user demographics and platform culture make it exceptionally effective for trust-building activities that support higher-value service sales and long-term business relationships. The platform's users tend to be older, more established, and more willing to invest in premium services when they trust the provider's expertise and credibility.

The trust-building process on Facebook operates differently than on entertainment-focused platforms. Facebook users expect more detailed information, appreciate comprehensive expertise demonstration, and value authentic relationship building over viral content or trending participation.

Developing Facebook as your primary trust-building platform requires consistent demonstration of expertise, transparent sharing of your business processes and results, genuine engagement with your audience's questions and concerns, and systematic nurturing of relationships over extended periods.

Content Strategy for Trust Building

Trust-building content on Facebook should be comprehensive, valuable, and authentic rather than entertaining or viral-focused. Share detailed insights about your industry, comprehensive tutorials that demonstrate your expertise, behind-the-scenes content that shows your work process, and case studies that highlight your results.

Long-form content performs well on Facebook because users expect more detailed information than they seek on other platforms. Use this preference for depth to share comprehensive guides, detailed analyses of industry trends, thorough explanations of your methodologies, and extensive case studies that showcase your expertise.

Personal storytelling builds trust by humanizing your expertise and helping potential clients understand your journey and motivations. Share challenges you've overcome, lessons you've learned, and the experiences that shaped your current expertise. This authentic sharing creates emotional connections that support business relationships.

Client success stories and testimonials should be featured prominently but authentically. Share detailed case studies that explain the client's challenge, your solution approach, the implementation process, and the achieved results. This comprehensive approach builds confidence in your abilities while providing valuable insights for other potential clients.

Authority Building Through Value Delivery

Consistent value delivery positions you as the go-to expert in your niche while building the trust necessary for high-value service sales. Focus on solving real problems for your audience through your content, even when you're not directly promoting your services.

Answer questions thoroughly in your Groups and on your Page, providing detailed, actionable advice that helps people achieve immediate improvements. This generous sharing builds reciprocity and positions you as someone who genuinely cares about helping others succeed.

Share industry insights and trends that help your audience stay informed about developments that affect their businesses or lives. Being a reliable source of important information builds trust and positions you as someone worth following and eventually hiring.

Create comprehensive resources like guides, templates, or frameworks that provide significant value while demonstrating your expertise. These resources serve as lead magnets while showcasing the quality and depth of your knowledge.

Relationship Nurturing Systems

Facebook's features support systematic relationship nurturing that can convert social media connections into long-term business relationships. Use these features strategically to maintain ongoing contact with potential clients and referral sources.

Regular engagement with your audience's content builds reciprocal relationships that can lead to business opportunities. Comment meaningfully on posts from potential clients, collaborators, and referral sources. This consistent presence keeps you top-of-mind when opportunities arise.

Use Facebook Messenger strategically to deepen relationships with high-value connections. Reach out personally to celebrate others' successes, offer help during challenges, or share resources that might benefit specific individuals. These personal touches build stronger relationships than public interactions alone.

Create systems for staying connected with former clients and satisfied customers who might become referral sources. Regular check-ins, sharing of relevant opportunities, and celebration of their successes helps maintain relationships that can generate ongoing business value.

MONETIZATION: FROM MEMBERSHIPS TO DIGITAL PRODUCTS

Facebook's established user base and sophisticated business tools make it an ideal platform for generating recurring revenue through memberships, subscriptions, and digital product sales. The platform's users are generally more comfortable with online purchases and recurring payments than users on newer social platforms.

The key to successful Facebook monetization is building comprehensive value delivery systems that justify ongoing financial relationships with your audience. This means creating memberships, courses, and digital products that provide continuous value rather than one-time benefits.

Facebook's integrated payment systems, group management tools, and advertising capabilities support sophisticated membership and product launch strategies that can generate substantial passive income when implemented systematically.

Membership Site Development

Facebook Groups can function as membership sites when combined with external payment processing and content delivery systems. Create private Groups for paying members that provide exclusive content, direct access to your expertise, and peer networking opportunities that justify membership fees.

Successful memberships require consistent value delivery through regular exclusive content, monthly group calls or Q&A sessions, member-only resources and templates, priority access to your services or products, and facilitated networking among members with similar goals or challenges.

Price membership levels strategically based on the value provided and your target audience's budget capabilities. Consider multiple membership tiers that provide different levels of access and value, allowing members to choose investment levels that match their needs and financial situations.

Retention strategies are crucial for membership success because recurring revenue depends on ongoing member satisfaction. Regular surveys, responsive member support, continuous content improvement, and community-building activities help maintain high retention rates that maximize lifetime customer value.

Digital Product Launch Strategies

Facebook's advertising system and organic reach capabilities make it excellent for launching digital products like courses, ebooks, templates, and software tools. The platform's detailed targeting allows you to reach people most likely to purchase your specific offerings.

Product launches on Facebook should follow proven sequences that build anticipation, demonstrate value, and create urgency for purchase decisions. Pre-launch content that teases product benefits, launch week content that showcases product features and testimonials, and post-launch content that addresses common questions and objections.

Use Facebook Live strategically during product launches to provide real-time interaction, answer questions about your offering, demonstrate product features or benefits, and create urgency through limited-time bonuses or pricing. Live interaction builds trust and can significantly increase conversion rates.

Customer testimonials and case studies become crucial during digital product launches because potential buyers can't physically examine your offering. Share detailed success stories from beta users, demonstrate specific outcomes achieved through your product, and address common concerns or objections through social proof.

Passive Income System Creation

Building passive income through Facebook requires creating systems that generate revenue with minimal ongoing time investment. This typically involves combining evergreen content, automated marketing sequences, and systematized customer onboarding processes.

Evergreen content marketing involves creating valuable content that remains relevant over time and continues attracting potential customers long after publication. Educational posts, comprehensive guides, and detailed tutorials can generate leads and sales for months or years with minimal additional effort.

Email list integration is crucial for Facebook passive income because it allows you to nurture relationships and promote offers outside of Facebook's algorithm limitations. Use Facebook to drive email signups, then use email sequences to nurture subscribers and promote your products or services systematically.

Affiliate marketing can provide additional passive income streams by promoting products and services that benefit your audience. Choose affiliate partnerships carefully to maintain trust and only promote offerings that provide genuine value to your community members.

Automated sales systems can handle product sales, customer onboarding, and basic customer service without requiring your direct involvement. These systems allow your Facebook marketing efforts to generate revenue even when you're not actively managing your accounts or creating new content.

Action Steps for Chapter 7:

Action Item	Task Description	Expected Outcome
Facebook Business Infrastructure	Set up a professional Facebook Page with complete business information, create or optimize a Facebook Group around your expertise, and install Facebook Pixel on your website for advertising preparation.	Complete Facebook business presence ready for systematic growth and monetization
Trust-Building Content Calendar	Develop 30 days of trust-building content including case studies, detailed tutorials, behind-the-scenes insights, and client success stories. Focus on comprehensive value delivery rather than viral content.	Systematic approach to building authority and trust with Facebook's business-minded audience
Group Engagement Strategy	Create daily engagement routines for your Facebook Group including valuable posts, member question responses, and community facilitation. Develop systems for identifying and nurturing potential customers within your group.	Active, engaged community that views you as the go-to expert and trusts your recommendations
Advertising Campaign Setup	Create your first Facebook ad campaign with clear objectives (lead generation, website traffic, or sales). Start with a small budget to test targeting and messaging before scaling successful campaigns.	Working knowledge of Facebook advertising with optimized campaigns driving business results
Passive Income Product Development	Design and launch one passive income product (digital course, membership, or affiliate program) specifically for your Facebook audience. Create automated systems for sales and customer onboarding.	Recurring revenue stream that leverages your Facebook presence for systematic income generation

EIGHT

YouTube Authority

YouTube's dual content ecosystem presents creators with two distinct paths to building authority and generating revenue. Long-form content builds deep expertise demonstration and sustained audience engagement, while YouTube Shorts provides rapid discovery and audience acquisition opportunities similar to TikTok's short-form format.

The most successful YouTube creators understand that long-form and Shorts content serve different strategic purposes and should be integrated rather than treated as competing formats. Long-form videos establish expertise and build trust through comprehensive coverage of topics, while Shorts attract new subscribers and drive traffic to your comprehensive content.

This strategic integration allows creators to capture audiences with different consumption preferences while building authority across the platform's entire ecosystem. Viewers who discover you through Shorts might become long-form subscribers, and long-form viewers might share your Shorts to expand your reach.

Long-form Content Authority Building

Long-form YouTube videos (typically 8+ minutes) provide opportunities to demonstrate comprehensive expertise, build trust through detailed explanation, and create valuable resources that viewers return to repeatedly. This format excels at establishing you as a definitive authority in your niche.

Educational long-form content should be structured like valuable courses or workshops, providing step-by-step guidance that viewers can implement immediately. Break complex topics into clear sections, use visual aids to support your explanations, and provide timestamps that allow viewers to navigate easily to relevant sections.

The depth possible in long-form content allows you to address complex topics that shorter formats cannot cover adequately. This comprehensive coverage builds authority because viewers recognize the thoroughness of your knowledge and your commitment to helping them understand difficult concepts.

Long-form content also supports better monetization through mid-roll ads, detailed product demonstrations, comprehensive affiliate reviews, and in-depth case studies that build confidence in your recommendations. The extended time allows for natural integration of monetization elements without compromising content quality.

YouTube Shorts Growth Acceleration

YouTube Shorts leverages the platform's massive user base to provide rapid audience growth opportunities for creators who understand the format's unique requirements. Shorts appear in dedicated feeds, suggested content areas, and search results, providing multiple discovery pathways.

Effective Shorts strategy involves adapting your long-form expertise into bite-sized, engaging pieces that create curiosity about your comprehensive content. Use Shorts to share quick tips, highlight key insights from longer videos, or provide previews of detailed tutorials available in your long-form content.

Shorts optimization requires understanding mobile-first consumption patterns and vertical video preferences. Create visually engaging content that works well on mobile devices, use clear text overlays for accessibility, and ensure your message is clear within the first few seconds to capture viewer attention.

The algorithm for Shorts differs from long-form content, prioritizing engagement velocity and completion rates over watch time duration. Focus on creating compelling hooks, maintaining visual interest throughout, and ending with strong calls-to-action that drive channel subscriptions or long-form video views.

Integration Strategy for Maximum Impact

The most effective YouTube strategy uses Shorts and long-form content synergistically to maximize both discovery and authority building. Create Shorts that tease insights from your long-form videos, driving traffic to comprehensive content while serving standalone value.

Repurpose your best long-form content into multiple Shorts by extracting key insights, memorable quotes, or actionable tips. This approach multiplies the value of your long-form creation time while providing consistent Short-form content for algorithm feeding.

Use your long-form content to provide context and depth for topics introduced in Shorts. When Shorts viewers want more detailed information, they can easily find comprehensive coverage on your channel, creating a natural progression from casual viewer to dedicated subscriber.

Cross-promote strategically by mentioning your Shorts in long-form videos and referring Shorts viewers to specific long-form content for deeper learning. This internal linking builds watch time across your entire channel while helping viewers find the content format they prefer.

YOUTUBE SEO & RANKING HACKS

YouTube functions as the world's second-largest search engine, making SEO optimization crucial for long-term channel success. Unlike social media platforms that prioritize recent content, YouTube's search-driven discovery means properly optimized videos can attract viewers for years after publication.

YouTube SEO operates on multiple ranking factors including video title optimization, description keyword integration, tag relevance, thumbnail click-through rates, audience retention metrics, and engagement signals. Understanding these factors allows creators to optimize content for both immediate performance and long-term discoverability.

The platform's sophisticated algorithm considers user behavior signals heavily, meaning that SEO optimization must be balanced with genuine audience value. Videos that are optimized for search but fail to satisfy viewer intent will struggle with retention metrics that ultimately hurt their ranking performance.

Keyword Research and Title Optimization

YouTube keyword research begins with understanding what your target audience searches for when seeking information in your niche. Use YouTube's search suggestions, Google Keyword Planner, and tools like TubeBuddy or VidIQ to identify high-volume, low-competition keyword opportunities.

Effective titles balance keyword optimization with click-through appeal. Include your primary keyword early in the title while creating curiosity or promising specific benefits that encourage clicks. "How to Start a Profitable Blog in 2025 (Step-by-Step Beginner Guide)" works better than "Blogging Tutorial" because it's specific, includes keywords, and promises comprehensive coverage.

Long-tail keywords often provide better opportunities for smaller channels because they face less competition while attracting more qualified viewers. Instead of targeting "fitness tips," consider "fitness tips for busy moms over 40," which attracts a specific audience more likely to engage deeply with relevant content.

Title testing can improve performance over time. YouTube allows title changes after publication, so monitor your click-through rates and test different variations to optimize performance. Sometimes small changes in wording can significantly impact discoverability and engagement.

Description and Tag Optimization

YouTube video descriptions provide extensive opportunities for keyword integration and additional context that supports both SEO and viewer experience. The first 125 characters appear in search results and suggested videos, making this opening crucial for click-through optimization.

Structure descriptions with multiple sections: a compelling opening summary, a detailed video outline with timestamps, relevant keyword integration, links to related resources, and social media or website connections. This comprehensive approach serves both algorithmic requirements and viewer needs.

Tags help YouTube understand your video's topic and context, influencing which related videos and search results include your content. Use a mix of broad category tags, specific long-tail phrases, and branded tags that connect to your other content. Focus on relevance over volume; 10 relevant tags perform better than 20 loosely related ones.

Include semantic keywords and related terms throughout your description to help YouTube understand your content's comprehensive topic coverage. If your main keyword is "social media marketing," include related terms like "digital marketing," "online promotion," "social strategy," and platform-specific terms.

Advanced SEO Strategies

Video transcriptions and closed captions provide additional keyword optimization opportunities while improving accessibility. YouTube can read caption text, so including relevant keywords naturally in your spoken content boosts SEO while serving hearing-impaired viewers.

Playlist optimization helps establish topical authority by grouping related videos together. Create playlists around specific keyword themes, optimize playlist titles and descriptions, and organize content logically to encourage binge-watching that boosts overall channel metrics.

End screen and card optimization can improve session duration by guiding viewers to related content on your channel. Strategic linking to relevant videos keeps viewers engaged with your content longer, sending positive signals to YouTube's algorithm about your content's value.

Community engagement signals like comments, likes, and shares influence ranking performance. Encourage engagement through specific questions, respond to comments promptly to encourage discussion, and create content that naturally prompts viewer interaction and sharing.

Technical Optimization Elements

Thumbnail optimization significantly impacts click-through rates, which directly influence ranking performance. Create custom thumbnails that are visually appealing, clearly represent your content, include readable text, and stand out in suggested video feeds.

Upload timing and consistency signal channel reliability to YouTube's algorithm. Establish regular posting schedules and maintain consistency to build audience expectations and algorithmic favor. Channels that post consistently often receive better distribution than sporadic uploaders.

Video quality factors, including resolution, audio clarity, and production value influence viewer retention and satisfaction. While content quality matters most, technical proficiency helps maintain viewer engagement and supports positive algorithmic signals.

Engagement optimization involves structuring videos to maintain viewer attention throughout. Use pattern interrupts, visual variety, clear transitions, and compelling content pacing to maximize average view duration, which is a crucial ranking factor for YouTube's algorithm.

MONETIZATION: ADS, SPONSORSHIPS, AFFILIATE INCOME

YouTube offers more diverse monetization opportunities than most social platforms, allowing creators to generate revenue through platform-native advertising, direct sponsorship deals, affiliate marketing, product sales, and premium content offerings. Building sustainable YouTube income requires understanding and optimizing multiple revenue streams simultaneously.

The platform's monetization ecosystem rewards creators who build engaged, loyal audiences rather than those who simply accumulate high view counts. Advertisers and sponsors value creators who can influence purchasing decisions and drive meaningful engagement with promoted products or services.

Successful YouTube monetization requires balancing audience value with revenue generation. Overly promotional content alienates viewers and hurts long-term growth, while under-monetized channels miss opportunities to build sustainable creator businesses that support continued content creation.

YouTube Partner Program Optimization

YouTube's Partner Program provides ad revenue sharing that can generate significant income for channels with substantial viewership. To qualify, channels need 1,000 subscribers and 4,000 watch hours in the past 12 months, making subscriber growth and content consistency crucial for monetization eligibility.

Ad revenue optimization involves understanding YouTube's advertiser-friendly content guidelines and creating content that attracts premium advertising. Educational, family-friendly, and professionally produced content typically generates higher ad rates than controversial or niche content.

Video length affects ad revenue potential because longer videos can include multiple ad placements. Videos over 8 minutes can include mid-roll ads in addition to pre-roll and post-roll advertisements, potentially doubling or tripling ad revenue compared to shorter content.

Audience demographics significantly impact ad revenue because advertisers pay different rates for different audience segments. Viewers from high-income countries, professional demographics, and purchase-intent categories generate higher ad revenue than other audience segments.

Sponsorship and Brand Deal Strategies

Direct sponsorships often generate more revenue than ad sharing because brands pay premium rates for authentic creator endorsements. Building sponsorship opportunities requires demonstrating audience engagement, maintaining professional presentation, and developing expertise in sponsor-relevant topics.

Successful sponsorship content feels natural and valuable rather than purely promotional. The best creator sponsorships provide genuine product reviews, demonstrate real use cases, or integrate sponsored messages into educational content that serves the audience's interests.

Media kit development helps attract and close sponsorship deals by providing potential sponsors with audience demographics, engagement statistics, previous sponsorship examples, and clear pricing information. Professional presentation builds confidence in your ability to represent brands effectively.

Long-term sponsorship relationships often provide better revenue stability than one-off deals. Focus on building ongoing relationships with brands that align with your content and audience values, creating mutually beneficial partnerships that support consistent income.

Affiliate Marketing Integration

Affiliate marketing allows creators to earn commissions by promoting products and services they genuinely use and recommend. Successful affiliate integration requires choosing products that provide real value to your audience while generating meaningful revenue for your business.

Product demonstration and tutorial content works exceptionally well for affiliate promotion because it shows products in use while providing educational value. Detailed reviews, comparison videos, and tutorial content that features affiliate products can generate substantial commission income.

Disclosure and authenticity are crucial for affiliate success and legal compliance. Always disclose affiliate relationships clearly, only promote products you genuinely believe in, and maintain honest perspectives in your reviews and recommendations to preserve audience trust.

Affiliate income tracking and optimization involves monitoring which products generate the best conversion rates and focusing promotion efforts on high-performing affiliate relationships. Use analytics to understand which content types drive affiliate sales most effectively.

Premium Content and Product Sales

YouTube can drive traffic to premium content offerings like online courses, coaching programs, consulting services, and digital products. The platform's authority-building capabilities make it excellent for promoting high-value services and educational products.

Course creation and promotion through YouTube involves using free content to demonstrate teaching ability and subject expertise while promoting comprehensive paid programs for deeper learning. This approach provides immediate value while building desire for premium offerings.

Merchandise and physical product sales can be integrated into YouTube content through product placements, behind-the-scenes content, and lifestyle integration that showcases products naturally. YouTube's shopping features make it easier to convert viewers into customers directly through the platform.

Service-based business promotion uses YouTube content to demonstrate expertise, share case studies, and build trust with potential clients. Professional services, consulting, and coaching businesses can generate substantial revenue by using YouTube as a client acquisition and authority-building platform.

Action Steps for Chapter 8:

Action Item	Task Description	Expected Outcome
Content Strategy Integration	Plan 20 pieces of content that work as both long-form videos and Shorts series. Create comprehensive tutorials that can be broken into multiple Shorts while maintaining value in both formats.	Efficient content creation system that maximizes reach across YouTube's entire ecosystem
SEO Optimization Audit	Research 50 keywords in your niche using YouTube's search suggestions and SEO tools. Optimize your last 10 video titles, descriptions, and tags based on keyword research while maintaining natural readability.	Improved search visibility and long-term organic traffic growth from optimized content

Action Item	Task Description	Expected Outcome
Monetization Channel Setup	Apply for YouTube Partner Program if eligible, or create a plan to reach eligibility requirements. Set up affiliate accounts for relevant products and create a media kit for potential sponsors.	Multiple revenue streams ready for activation as your channel grows
Authority Content Series	Plan and outline a comprehensive video series (8–12 episodes) that establishes your expertise in your core topic. Create content that serves as a definitive resource viewers will return to and share.	Authoritative content library that builds trust and positions you as a go-to expert
Performance Tracking System	Set up YouTube Analytics monitoring for key metrics including watch time, click-through rates, subscriber conversion, and revenue performance. Create monthly review process for optimization insights.	Data-driven approach to improving content performance and monetization effectiveness

NINE

Content that Sells (Without Being Salesy)

Creating content that drives sales without feeling promotional requires understanding the psychological triggers that influence human decision-making and purchasing behavior. The most effective selling happens when audiences don't realize they're being sold to. They simply feel compelled to take action because your content has created genuine desire and addressed real needs.

Modern consumers are highly resistant to direct sales messages, having developed sophisticated filters for obvious promotional content. However, they remain highly responsive to content that educates, entertains, or inspires while subtly demonstrating value and building trust. This shift requires creators to master indirect influence rather than direct persuasion.

The psychology of social media selling operates on relationship building, value demonstration, and trust development rather than traditional sales techniques. When audiences trust your expertise and believe in your genuine desire to help them succeed, they naturally seek out your products and services as logical next steps in their journey.

The Trust-First Selling Framework

Trust forms the foundation of all successful social media selling because online audiences must believe in both your expertise and your intentions before they'll consider purchasing from you. Building this trust requires consistent value delivery over time rather than immediate sales pushes.

Demonstrate expertise through helpful content that solves real problems, shares valuable insights, and provides actionable advice that audiences can implement immediately. Each piece of valuable content makes a deposit in your trust account that supports future sales conversations.

Authenticity accelerates trust building by allowing audiences to connect with your genuine personality and motivations. Share your real experiences, admit mistakes and lessons learned, and show the human side of your expertise rather than presenting a polished, perfect image.

Social proof reinforces trust by showing that others have succeeded with your help. Client testimonials, case studies, transformation stories, and user-generated content all demonstrate that your methods work and that you deliver on your promises.

Emotional Triggers in Content Creation

Effective selling content connects with audiences on emotional levels before presenting logical reasons for purchase. Emotions drive purchasing decisions, while logic provides justification for choices already made emotionally.

Pain point activation involves identifying and articulating the specific frustrations, challenges, or desires that your audience experiences. When you accurately describe their problems, audiences immediately recognize your understanding and become more receptive to your solutions.

Aspiration building creates emotional desire by helping audiences envision their improved future after using your products or services. Paint vivid pictures of their transformed lives, enhanced capabilities, or achieved goals to create motivation for taking action.

Key Psychological Triggers for Social Media Selling:

- **Reciprocity Activation**: Provide significant value through free content, resources, or advice that creates natural desire to return the favor through purchases or referrals
- **Authority Demonstration**: Share credentials, results, media mentions, or expert recognition that positions you as a credible source worth trusting and following
- **Social Validation**: Showcase community engagement, customer testimonials, follower interactions, and peer endorsements that prove others value your expertise
- **Scarcity Creation**: Introduce limited availability, time-sensitive offers, or exclusive access that motivates timely decision-making without feeling manipulative
- **Consistency Leverage**: Help audiences make small commitments through engagement, email signups, or content sharing that creates momentum toward larger purchases
- **Liking Enhancement**: Build personal connections through shared values, similar experiences, humor, or personality traits that make audiences genuinely enjoy your content

The Value-First Sales Sequence

Structure your content to provide immediate value while naturally leading audiences toward your offerings. This sequence builds trust and desire simultaneously without triggering sales resistance.

Start by identifying problems your audience faces, then provide partial solutions through your content while hinting at more comprehensive solutions available through your products or services. This approach demonstrates your ability to help while creating desire for complete solutions.

Share case studies and success stories that show your methods working for others. These stories provide social proof while subtly demonstrating the transformation possible through your offerings.

Address objections and concerns preemptively through your content. When audiences see that you understand their hesitations and can address them thoughtfully, they become more confident in your ability to help them succeed.

STORYTELLING FRAMEWORKS FOR SOCIAL MEDIA

Stories are the most powerful tool for selling without being salesy because they engage emotions, create connections, and allow audiences to envision themselves in similar situations. Effective storytelling transforms abstract benefits into concrete experiences that audiences can relate to and desire for themselves.

The human brain is wired to respond to stories more than facts or features. Stories create emotional engagement, improve memory retention, and help audiences visualize outcomes in ways that straight information cannot achieve. This makes storytelling essential for social media content that needs to capture attention and drive action quickly.

Social media storytelling requires adapting traditional narrative structures to shorter attention spans and mobile consumption patterns. Your stories need to hook audiences immediately, deliver value quickly, and create clear connections between the story and your audience's needs or desires.

The Transformation Story Framework

Transformation stories follow a predictable structure that audiences find compelling: beginning state (problem/challenge), catalyst event (decision to change), struggle period (obstacles and efforts), breakthrough moment (solution discovery), and end state (achieved transformation). This structure works because it mirrors the journey your audience wants to experience.

Use transformation stories to showcase client successes, your own journey of overcoming challenges, or case studies of your methods working in real situations. These stories provide social proof while allowing audiences to envision similar transformations in their own lives.

Personal transformation stories are particularly powerful because they demonstrate that you've faced similar challenges to your audience and successfully overcome them. This creates immediate connection and credibility while showing that transformation is possible.

Client transformation stories showcase your ability to help others achieve desired outcomes. Focus on specific, measurable results and include emotional elements that help audiences connect with the client's experience and desired outcomes.

The Problem-Agitation-Solution Method

This storytelling approach identifies problems your audience faces, agitates those problems by highlighting their consequences, then presents your solution as the logical resolution. When done skillfully, this method feels helpful rather than manipulative.

Problem identification requires deep understanding of your audience's specific challenges, frustrations, and pain points. The more accurately you describe their problems, the more connected they'll feel to your content and solutions.

Agitation involves exploring the consequences of not addressing the problem—missed opportunities, continued frustration, wasted time or money, or failure to achieve important goals. This creates urgency for finding solutions.

Solution presentation should feel natural and logical given the problem and consequences you've established. Your offering becomes the obvious choice for addressing the challenges you've described.

The Hero's Journey for Business Content

Adapt the classic hero's journey narrative structure for business-focused content by positioning your audience as the hero and yourself as the guide who helps them overcome challenges and achieve their goals.

Present your audience as capable individuals facing legitimate challenges that require new knowledge, skills, or tools to overcome. This positioning respects their intelligence while acknowledging their need for assistance.

Position yourself as the experienced guide who has helped others overcome similar challenges. Share your credentials, experience, and track record of success without making yourself the hero of the story.

Provide the tools, knowledge, or support that enables your audience to succeed in their journey. Your products or services become the magical weapons or special knowledge that helps heroes achieve their quests.

Micro-Storytelling for Social Media

Social media requires condensed storytelling that delivers impact quickly. Master micro-storytelling by focusing on single moments, specific details, and emotional peaks rather than comprehensive narratives.

Use vivid details and specific language to make short stories feel real and relatable. Instead of "I had a difficult client," try "My client called at 11 PM demanding a complete strategy revision for their 8 AM presentation."

Focus on emotional moments that create connection—moments of doubt, breakthrough realizations, unexpected challenges, or surprising successes. These emotional peaks make stories memorable and shareable.

Connect every story to your audience's experience by explicitly drawing parallels between your story and their likely situations. Help them see themselves in your narratives and envision similar outcomes.

CTA (CALL-TO-ACTION) FORMULAS THAT CONVERT

Effective calls-to-action guide audiences toward desired actions without creating sales pressure or resistance. The most successful CTAs feel like natural next steps in the relationship rather than pushy sales attempts. They provide clear value propositions while making it easy for interested audiences to take action.

CTA effectiveness depends on timing, positioning, value clarity, and friction reduction. The right CTA at the right moment can dramatically increase conversion rates, while poorly timed or unclear CTAs can damage trust and reduce engagement.

Different types of content require different CTA approaches. Educational content might call for engagement or deeper learning, behind-the-scenes content might encourage following or community joining, and testimonial content might prompt direct inquiry about services.

The Soft CTA Strategy

Soft CTAs encourage action without creating sales pressure by focusing on value delivery rather than purchase requests. These approaches build trust while moving audiences through your sales funnel gradually.

Educational CTAs invite audiences to learn more about topics you've introduced in your content. "Want to learn the complete framework I use with my clients?" is more appealing than "Buy my course now" because it focuses on learning rather than purchasing.

Community CTAs invite audiences to join conversations, groups, or communities where they can connect with like-minded people and continue learning. This approach builds relationships while positioning you as a community leader.

Resource CTAs offer valuable free resources that require email signup or profile following. Lead magnets, templates, checklists, or guides provide immediate value while building your email list for future marketing.

The Problem-Solution CTA Bridge

Connect your content's problem identification directly to your CTA by positioning your offering as the logical next step for solving the challenges you've discussed.

After presenting problems and their consequences in your content, your CTA becomes the natural solution path: "If you're struggling with [specific problem], here's how I can help..." This connection feels logical rather than promotional.

Use bridge language that connects your content to your CTA smoothly. Phrases like "The next step is..." or "If you want to solve this faster..." create natural transitions from content to action requests.

Provide multiple action options to accommodate different readiness levels. Some audiences may be ready for direct purchases, while others prefer email signup or consultation scheduling. Offering options increases overall conversion rates.

The Curiosity Gap CTA

Create curiosity gaps that can only be satisfied by taking your desired action. This approach leverages psychological principles that make completion of started tasks feel compelling.

Tease valuable information available through your CTA without revealing everything in your current content. "The three-step framework I mentioned is available in my free guide" creates desire to access the complete information.

Use incomplete information strategically to encourage CTA engagement. Share part of a process, formula, or strategy while making the complete version available through your desired action.

Promise exclusive or advanced content available only to email subscribers, group members, or product customers. This exclusivity creates desire while rewarding people who take your suggested actions.

Platform-Specific CTA Optimization

Each social media platform has different user expectations and optimal CTA approaches. Tailor your calls-to-action to match platform culture and user behavior patterns.

Instagram CTAs work well when integrated into Stories, highlighted in bio links, or included naturally in post captions. The visual nature of the platform supports CTAs that promise visual resources or behind-the-scenes access.

TikTok CTAs should be quick and easy to remember since users often scroll quickly. Simple actions like "Follow for more tips" or "Link in bio for the full guide" work better than complex instructions.

YouTube CTAs can be more detailed because users expect longer content and have more attention bandwidth. End-screen CTAs, description links, and verbal requests during videos all perform well.

Facebook CTAs benefit from the platform's trust-building culture. Invitation-based language works better than demand-based requests. "I'd love to help you with this" performs better than "You need to buy this."

CTA Testing and Optimization

Continuously test different CTA approaches to identify what resonates most with your specific audience. Small changes in wording, timing, or positioning can significantly impact conversion rates.

A/B test CTA placement within your content. Some audiences respond better to early CTAs that capture immediate interest, while others prefer CTAs after receiving substantial value from your content.

Test different value propositions in your CTAs. Some audiences respond to time-saving benefits, others to money-making opportunities, and still others to skill development or personal growth promises.

Monitor engagement metrics to understand which CTAs generate the highest conversion rates and lifetime customer value. Focus your optimization efforts on CTAs that not only convert but also attract your ideal customers.

Action Steps for Chapter 9:

Action Item	Task Description	Expected Outcome
Psychology Audit	Analyze your last 20 pieces of content to identify which psychological triggers you're currently using. Map each post to specific emotional outcomes and trust-building elements, then identify gaps to address.	Clear understanding of your current influence strategies and opportunities for psychological optimization
Story Bank Creation	Develop 15 transformation stories from your personal experience and client work. Create templates for different story types including challenge-to-success, mistake-to-lesson, and discovery-to-implementation narratives.	Ready-to-use story collection that demonstrates value and builds emotional connection with audiences

Action Item	Task Description	Expected Outcome
CTA Formula Testing	Create and test 5 different CTA approaches across your platforms. Compare soft CTAs, curiosity-gap CTAs, and problem-solution bridges to identify which generates highest engagement and conversion for your audience.	Optimized call-to-action system that maximizes conversion without damaging audience trust
Value-First Content Calendar	Plan 30 days of content using the value-first selling approach. Each piece should provide substantial value while naturally leading toward your offerings without direct promotion.	Systematic approach to selling through service that builds trust while generating business results
Conversion Tracking Setup	Implement tracking systems to measure how your content leads to email signups, website visits, and sales. Create attribution systems that connect social media engagement to business outcomes.	Data-driven understanding of which content types and selling approaches generate the best business results

TEN

Building Multiple Income Streams

Affiliate marketing represents one of the fastest paths to monetizing social media influence because it requires no product creation, inventory management, or customer service responsibilities. You simply recommend products and services you already use and believe in, earning commissions when your audience makes purchases through your unique links.

The key to successful affiliate marketing lies in authentic recommendation rather than aggressive promotion. Your audience trusts your judgment and looks to you for honest guidance about tools, products, and services that can help them achieve their goals. This trust becomes valuable when channeled toward high-quality affiliate partnerships that genuinely benefit your followers.

Building sustainable affiliate income requires strategic partner selection, authentic integration methods, and systematic promotion approaches that maintain audience trust while generating meaningful revenue. The most successful affiliate marketers treat their recommendations as extension of their expertise rather than separate income-generating activities.

Strategic Affiliate Partner Selection

Choose affiliate partnerships based on your audience's genuine needs and your personal experience with products rather than commission rates alone. High-converting affiliate content comes from authentic enthusiasm and detailed knowledge that can only result from real product usage.

Focus on affiliate partners that align with your content themes and audience demographics. A productivity coach should promote productivity tools, project management software, and business education rather than random high-commission products that don't serve their audience's specific needs.

Research potential affiliate programs thoroughly by examining commission structures, payment terms, promotional restrictions, and brand reputation. Some programs offer higher commissions but have poor conversion rates or restrictive promotional guidelines that limit your earning potential.

Test products personally before promoting them to ensure they meet your quality standards and deliver promised benefits. Your reputation depends on the quality of your recommendations, so only promote products you would recommend even without earning commissions.

Content Integration Methods

Integrate affiliate promotions naturally into your existing content rather than creating obviously promotional posts that can damage audience trust and engagement. The most effective affiliate content provides value first while subtly incorporating product recommendations.

Tutorial and educational content works exceptionally well for affiliate promotion because it demonstrates products in use while providing genuine value. Show your audience how to achieve specific outcomes using affiliate products, making the recommendations feel helpful rather than sales-focused.

Review and comparison content allows detailed discussion of affiliate products' benefits, drawbacks, and use cases. Honest reviews that acknowledge both strengths and limitations build trust while helping audiences make informed purchasing decisions.

Behind-the-scenes content can showcase affiliate products as part of your regular workflow or business operations. When audiences see products integrated into your real work processes, recommendations feel authentic and trustworthy.

Advanced Affiliate Marketing Techniques:

1. **Email Sequence Integration**: Create email sequences that nurture affiliate relationships over time rather than relying on single promotional messages. Share case studies, tips for using affiliate products, and exclusive bonuses that add value while encouraging purchases through your links.
2. **Bundle Creation and Bonuses**: Negotiate with affiliate partners to create exclusive bonuses for your audience, such as additional training, templates, or consultation time. These bonuses increase perceived value and differentiate your promotions from other affiliates promoting the same products.
3. **Long-tail Content Strategy**: Create comprehensive content around specific use cases for affiliate products rather than general promotional posts. "How I Use [Product] to Save 10 Hours Per Week" performs better than "Why You Should Buy [Product]" because it provides specific value while demonstrating product benefits.
4. **Cross-Platform Promotion**: Develop affiliate promotion strategies that work across all your social media platforms, adapting the core message to each platform's unique culture and audience expectations while maintaining consistent recommendation quality.

CREATING DIGITAL PRODUCTS (COURSES, EBOOKS, COACHING)

Digital product creation transforms your expertise into scalable income streams that can generate revenue repeatedly without ongoing time investment per sale. Unlike service-based offerings that trade time for money, digital products allow you to package your knowledge once and sell it indefinitely.

The digital product landscape includes online courses, ebooks, templates, software tools, membership sites, and coaching programs. Each format serves different audience preferences and learning styles while providing various levels of interaction and price points.

Successful digital product creation requires understanding your audience's specific challenges, preferred learning formats, and willingness to invest in solutions. The most profitable digital products solve urgent problems for audiences who have both motivation and budget to invest in solutions.

Course Creation and Marketing

Online courses represent premium digital products that can generate substantial revenue while establishing your expertise and building lasting relationships with students. Courses work best for complex topics that benefit from structured learning progression and comprehensive coverage.

Effective course creation starts with identifying specific outcomes your audience wants to achieve and designing a curriculum that guides them from their current state to desired results. Focus on transformation rather than information—students buy courses to change their situation, not just learn facts.

Course content should balance theoretical knowledge with practical application through exercises, templates, case studies, and action steps that help students implement what they're learning. This practical focus improves student success rates while generating testimonials that support future marketing.

Marketing courses require building anticipation before launch through free content that demonstrates your teaching ability, sharing student success stories, and providing previews of course content that create desire without giving away everything for free.

Ebook and Guide Development

Ebooks and digital guides serve audiences who prefer self-paced reading and reference materials they can revisit repeatedly. These products work well for step-by-step processes, comprehensive frameworks, and detailed reference materials.

Focus ebook topics on specific, actionable subjects that audiences can implement immediately after reading. "The Complete Guide to Instagram Growth" provides more value than "Social Media Tips" because it promises comprehensive coverage of a specific outcome.

Ebook promotion benefits from content marketing that shares excerpts, key insights, and practical tips from your guide while positioning the complete ebook as essential for full implementation. This approach demonstrates value while creating desire for comprehensive coverage.

Price ebooks strategically based on length, depth, and audience budget capacity. Lower-priced ebooks can serve as entry-level products that introduce audiences to your work and build trust for higher-priced offerings later.

Coaching Program Development

Coaching programs combine your expertise with personalized guidance, creating premium offerings that can command high prices while delivering transformational results for clients. Coaching works best when you have proven methods and enjoy working directly with people.

Structure coaching programs with clear processes, expected outcomes, and defined timelines that help potential clients understand what they'll receive and achieve. Vague coaching promises are less compelling than specific transformation programs with measurable results.

Group coaching programs provide scalable alternatives to one-on-one coaching by serving multiple clients simultaneously while maintaining personalized attention and community benefits. This format increases your hourly earning potential while reducing client costs.

Coaching program marketing relies heavily on social proof, case studies, and demonstration of your coaching methodology through free content that showcases your ability to help others achieve results.

MEMBERSHIPS, SUBSCRIPTIONS, AND BRAND DEALS

Recurring revenue streams provide business stability and predictable income that supports long-term planning and growth investment. Memberships, subscriptions, and ongoing brand partnerships create compound income that grows over time rather than requiring constant new customer acquisition.

These revenue models work by providing ongoing value that justifies recurring payments. Members and subscribers continue paying because they receive consistent benefits that exceed the cost of their investment.

Building successful recurring revenue requires understanding what your audience values enough to pay for repeatedly and creating systems that deliver that value consistently without requiring proportional increases in your time investment.

Membership Site Strategies

Membership sites provide exclusive access to premium content, community interaction, and ongoing education in exchange for monthly or annual fees. Successful memberships feel like valuable clubs that members are proud to join and reluctant to leave.

Content planning for memberships should balance exclusive new material with evergreen resources that provide ongoing reference value. Members should feel they're receiving fresh value each month while building a library of useful resources over time.

Community features distinguish membership sites from simple content libraries by providing interaction with you and other members. Discussion forums, monthly live calls, and peer networking opportunities create additional value that justifies ongoing membership fees.

Member retention requires consistent value delivery, responsive communication, and continuous improvement based on member feedback. Track which content types generate the most engagement and satisfaction to optimize your ongoing content strategy.

Subscription-Based Content

Content subscriptions work for creators who can provide regular, high-value content that audiences want to receive consistently. Newsletter subscriptions, video series, and premium content access all support subscription models.

Subscription content should provide unique value that isn't available through your free channels. Exclusive insights, advanced strategies, behind-the-scenes access, or early content access can justify subscription fees.

Pricing subscriptions requires balancing affordability with value perception. Start with lower prices to build a subscriber base, then increase prices as you add more value and demonstrate consistent quality.

Subscription growth benefits from free trial periods, preview content that demonstrates subscription value, and word-of-mouth referrals from satisfied subscribers who share your content with others.

Brand Partnership Development

Long-term brand partnerships provide stable income streams while building professional relationships that can lead to additional opportunities. Focus on developing ongoing relationships rather than one-off sponsored posts.

Brand partnership success requires professional communication, consistent content quality, and measurable results that demonstrate your value to brand partners. Track engagement metrics, click-through rates, and conversion data to show partners the ROI of working with you.

Negotiate partnerships that align with your content themes and audience interests to maintain authenticity while generating revenue. Partnerships that feel natural to your audience perform better for both you and brand partners.

Diversify brand partnerships across multiple companies and industries to reduce dependence on single relationships while maximizing revenue opportunities from your influence and expertise.

Premium Community Development

Premium communities combine membership, content, and coaching elements to create comprehensive programs that command higher prices while delivering exceptional value. These communities work best for creators with strong personal brands and deep expertise.

Structure premium communities around specific outcomes or transformations that members want to achieve. Clear focus helps attract committed members who are willing to invest significantly in their success.

Premium community pricing should reflect the comprehensive value provided through exclusive content, personal access, community networking, and outcome-focused programming. These programs often command $100-500+ monthly fees when positioned correctly.

Community management requires dedicated time and attention to maintain engagement, provide valuable content, and facilitate meaningful connections between members. Consider this time investment when pricing and structuring your community offerings.

Action Steps for Chapter 10:

Action Item	Task Description	Expected Outcome
Affiliate Program Research	Identify and apply to 5–7 affiliate programs that align with your niche and audience needs. Research commission structures, promotional guidelines, and product quality to ensure good matches for authentic promotion.	Strategic affiliate partnerships ready for authentic integration into your content strategy
Digital Product Planning	Choose one digital product format (course, ebook, or coaching) and create a detailed outline including target audience, learning outcomes, content structure, and pricing strategy. Validate the concept with your audience before full development.	Clear roadmap for creating your first scalable digital product that addresses genuine audience needs
Recurring Revenue Strategy	Design a membership or subscription offering that provides ongoing value to your audience. Plan monthly content themes, community features, and retention strategies that justify recurring payments.	Foundation for building predictable monthly income through value-driven recurring revenue
Income Stream Integration	Create a comprehensive monetization strategy that combines affiliate marketing, digital products, and recurring revenue in a way that feels natural and serves your audience effectively without overwhelming them with promotional content.	Diversified income portfolio that maximizes revenue opportunities while maintaining audience trust

Action Item	Task Description	Expected Outcome
Revenue Tracking System	Implement tracking systems for all income streams including affiliate commissions, digital product sales, and recurring revenue. Create monthly reporting that shows revenue trends and identifies optimization opportunities.	Data-driven understanding of which income streams perform best and deserve more focus

ELEVEN

Automation & Scaling With AI Tools

Automation transforms social media from a time-consuming daily grind into a systematic business operation that runs efficiently while you focus on high-value activities like strategy development, relationship building, and product creation. The key to successful automation is maintaining authenticity while gaining efficiency—your automated systems should feel human and responsive rather than robotic and impersonal.

Smart automation handles repetitive tasks that don't require human creativity or judgment while preserving your personal touch for interactions that build relationships and drive business results. This strategic approach allows you to maintain a consistent presence across multiple platforms without sacrificing quality or authentic connection with your audience.

The most successful creators use automation as a force multiplier that amplifies their human efforts rather than replacing them. Automated systems handle the mechanics of posting and basic engagement, while human oversight manages strategy, creative decisions, and meaningful relationship building.

Content Scheduling and Distribution

Content scheduling tools allow you to maintain consistent posting across multiple platforms without being chained to your devices. Tools like Buffer, Hootsuite, Later, and Creator Studio enable batch content creation during focused work sessions followed by automatic distribution according to optimal timing schedules.

Effective scheduling requires understanding each platform's optimal posting times for your specific audience and content types. Use platform analytics to identify when your followers are most active and engaged, then schedule content distribution to maximize visibility and engagement potential.

Cross-platform content adaptation through scheduling tools saves significant time by allowing you to create content once and automatically format it appropriately for different platforms. However, maintain platform-specific customization for captions, hashtags, and visual formats to ensure content feels native to each environment.

Content calendar planning becomes more efficient with scheduling automation. Plan themes, campaigns, and content series weeks or months in advance, then use automation tools to execute your strategy consistently while you focus on creating new content and engaging with your community.

Smart Engagement Automation

Engagement automation must be implemented carefully to maintain authenticity while managing the time demands of community interaction. Auto-responses, comment management, and initial outreach can be systematized without sacrificing personal connection quality.

Automated direct message responses can handle initial contact and frequently asked questions while ensuring that meaningful conversations are flagged for personal attention. Create templates for common inquiries while maintaining triggers that escalate complex or sales-related conversations to manual handling.

Comment management tools can filter spam, highlight priority comments, and suggest responses based on comment content and sentiment. However, maintain personal oversight for all public responses to ensure they align with your brand voice and provide appropriate value.

Social listening automation monitors mentions of your brand, relevant keywords, and industry conversations that present engagement opportunities. Set up alerts for important conversations where your input could provide value or build relationships with potential collaborators or customers.

Email Marketing Integration

Email automation integrates with your social media efforts to nurture relationships and drive conversions through systematic follow-up sequences. Social media attracts attention, while email automation converts that attention into lasting business relationships.

Lead magnet automation captures email addresses from social media traffic and immediately delivers promised resources while beginning nurture sequences designed to build trust and demonstrate value over time. This integration transforms social media followers into owned audience members you can reach directly.

Welcome sequences introduce new subscribers to your expertise, share your best content, and gradually present your products or services through value-first approaches. These automated sequences work continuously to convert new leads while you focus on creating fresh content and growing your audience.

Sales sequence automation follows up with prospects who have shown interest in your offerings but haven't yet purchased. These sequences provide additional value, address common objections, and present offers at appropriate intervals without requiring manual tracking and follow-up.

AI ANALYTICS FOR SMARTER DECISION-MAKING

AI-powered analytics transform overwhelming amounts of social media data into actionable insights that guide strategic decisions and optimize performance across all your platforms. Instead of manually analyzing metrics and guessing about audience preferences, AI analytics identify patterns and opportunities that would be impossible to detect through manual review.

Modern social media success requires understanding complex audience behaviors, content performance patterns, and engagement trends that span multiple platforms and time periods. AI analytics process this information continuously, providing real-time recommendations that help optimize your content strategy and resource allocation.

The most valuable AI analytics focus on predictive insights rather than just historical reporting. Understanding what content types are likely to perform well, when your audience will be most receptive to different messages, and which strategies will drive the best return on investment allows for proactive optimization rather than reactive adjustments.

Performance Pattern Recognition

AI analytics identify performance patterns across your content that reveal what drives engagement, conversions, and growth for your specific audience and niche. These insights help you double down on successful approaches while avoiding content types that consistently underperform.

Content performance analysis reveals which topics, formats, posting times, and engagement strategies generate the best results for your goals. AI can identify subtle patterns like seasonal content preferences, audience mood influences, and platform-specific optimization opportunities that manual analysis would miss.

Audience behavior insights help you understand how different segments of your following interact with your content, what drives them to take desired actions, and how to tailor content for maximum impact with each group. This segmentation allows for more targeted and effective content strategies.

Competitive analysis through AI tools reveals what's working for similar creators in your space, identifies content gaps you could fill, and highlights opportunities for differentiation that could give you competitive advantages.

Predictive Content Optimization

AI analytics predict which content types and topics are likely to perform well based on current trends, historical performance, and audience engagement patterns. This predictive capability allows you to create content with higher success probability rather than guessing about what will resonate.

Trending topic identification helps you capitalize on emerging conversations and viral opportunities before they become oversaturated. AI can identify rising topics in your niche while they still offer significant organic reach potential.

Optimal timing predictions consider audience activity patterns, platform algorithm preferences, and external factors like holidays or events that might influence engagement levels. This ensures your content reaches audiences when they're most likely to engage meaningfully.

Content gap analysis identifies topics and formats that your audience wants but aren't being adequately served by existing content. These gaps represent opportunities for high-performing content that meets unaddressed needs in your community.

ROI and Business Impact Measurement

AI analytics connect social media activities to actual business results by tracking how content performance translates into email signups, website traffic, product sales, and other meaningful business outcomes. This connection helps you focus efforts on activities that drive real business growth.

Attribution modeling shows which social media activities contribute to conversions throughout your sales funnel, allowing you to understand the true value of different content types and platforms rather than just measuring vanity metrics like likes and shares.

Customer lifetime value analysis connected to social media acquisition sources reveals which platforms and content types attract your most valuable customers. This insight guides resource allocation toward activities that generate the highest long-term business value.

Campaign effectiveness measurement shows which marketing campaigns, product launches, and promotional activities generate the best return on investment through social media channels. Use these insights to optimize future campaigns and promotional strategies.

SCALING FROM SIDE HUSTLE TO BUSINESS EMPIRE

Scaling a social media business requires transitioning from personal content creation to systematic business operations that can grow without proportional increases in your personal time investment. This transformation involves building teams, creating systems, and developing processes that maintain quality while increasing output and revenue.

The scaling process typically involves several stages: solo creator handling all activities, virtual assistant support for routine tasks, specialized team members for content creation and community management, and eventually full business operations with multiple revenue streams and team members handling most day-to-day activities.

Successful scaling requires maintaining the authentic personal brand that attracted your initial audience while building business infrastructure that can operate independently of your constant involvement. This balance allows for significant growth without losing the personal connection that drives social media success.

Team Building and Delegation

Building an effective team starts with identifying tasks that don't require your personal expertise or brand voice. Content research, graphic design, scheduling, basic customer service, and administrative tasks can often be delegated to team members while you focus on strategy and relationship building.

Virtual assistants can handle routine social media tasks like content scheduling, comment moderation, direct message responses, and basic graphic design. Choose assistants with social media experience and provide detailed training on your brand voice and community standards.

Content creators can help scale your content production by creating graphics, editing videos, writing captions, and developing content ideas under your guidance. Maintain quality control through review processes while allowing team members to handle execution.

Specialized roles like community managers, customer service representatives, and marketing coordinators become valuable as your business grows. These team members can handle specific aspects of your operation while you focus on high-level strategy and business development.

Systems and Process Development

Document all your successful processes to enable delegation and maintain consistency as your team grows. Create standard operating procedures for content creation, community engagement, customer service, and business operations that team members can follow independently.

Content production systems should include idea generation processes, creation workflows, approval procedures, and distribution schedules that ensure consistent quality output without requiring your direct involvement in every piece of content.

Customer journey mapping helps systematize how you move people from social media followers to paying customers through automated sequences, systematic follow-up processes, and clear conversion pathways that work without manual intervention.

Quality control systems ensure that delegated work maintains your brand standards through review processes, approval workflows, and feedback mechanisms that allow team members to improve while protecting your reputation.

Revenue Stream Diversification

Scaling requires developing multiple revenue streams that aren't dependent on your direct time investment. Digital products, affiliate marketing, advertising revenue, and licensing opportunities can generate income while you focus on business growth rather than direct service delivery.

Product development should focus on scalable offerings like online courses, membership sites, and digital tools that can serve unlimited customers without increasing your workload proportionally. These products provide leverage that allows significant income growth without proportional time increases.

Partnership development creates revenue opportunities through joint ventures, affiliate relationships, speaking engagements, and brand collaborations that leverage your expertise and audience while requiring minimal ongoing time investment.

Investment strategies involve reinvesting social media profits into assets, team members, and business systems that generate additional income streams beyond your primary social media activities. This diversification provides stability and growth potential beyond social media platforms.

Business Infrastructure Development

Legal and financial infrastructure becomes crucial as your social media business grows. Establish proper business entities, accounting systems, contract templates, and intellectual property protections that support professional business operations.

Technology systems should automate routine business operations like customer onboarding, product delivery, financial tracking, and team communication. Invest in tools and platforms that reduce manual work while improving service quality.

Strategic planning processes help guide business growth through goal setting, resource allocation, and opportunity evaluation. Regular strategic reviews ensure your business development aligns with your long-term objectives and market opportunities.

Exit strategy consideration involves building business systems and assets that have value independent of your personal involvement. This creates options for business sale, partnership opportunities, or passive income generation that provides long-term financial security.

Action Steps for Chapter 11:

Action Item	Task Description	Expected Outcome
Automation Audit and Setup	Identify 10 routine tasks in your social media workflow that can be automated. Implement scheduling tools, basic engagement automation, and email integration systems that save at least 10 hours per week.	Streamlined operations that free up time for high-value strategy and relationship building activities
AI Analytics Implementation	Set up AI-powered analytics tools (like Sprout Social, Hootsuite Insights, or similar) to track performance patterns, audience behavior, and predictive content opportunities across all your platforms.	Data-driven insights that improve content performance and business decision-making
Team Development Planning	Create job descriptions for your first virtual assistant or team member hire. Document your current processes and identify which tasks are ready for delegation while maintaining quality standards.	Clear roadmap for scaling beyond solo operations while maintaining brand quality and authenticity
Systems Documentation	Document your most successful processes for content creation, audience engagement, and customer conversion. Create standard operating procedures that enable team members to replicate your results.	Scalable business systems that allow growth without proportional increases in your personal time investment
Revenue Diversification Strategy	Develop plans for at least 3 scalable revenue streams that can operate with minimal ongoing time investment. Focus on digital products, automated services, or partnership opportunities that provide leverage.	Multiple income sources that reduce business risk while increasing profit potential through systematization

TWELVE

Future-Proofing Your Social Media Empire

Algorithm changes represent the greatest ongoing challenge for social media entrepreneurs because they can dramatically impact reach, engagement, and revenue overnight. However, creators who build their strategies on fundamental value delivery rather than algorithm manipulation consistently weather these changes while weaker competitors struggle to adapt.

The key to algorithm resilience lies in understanding that platforms fundamentally want to show users content that keeps them engaged and satisfied. While specific ranking factors may change, the core objective remains constant—create content that genuinely serves your audience's needs and interests, and you'll maintain algorithmic favor regardless of technical updates.

Building algorithm-resistant strategies requires focusing on engagement quality over quantity, authentic audience building over follower acquisition, and sustainable content approaches over viral tactics. These foundational principles remain valuable regardless of how platforms adjust their content distribution methods.

Understanding Platform Motivations

Every algorithm change serves the platform's business objectives, which typically involve increasing user engagement, time on platform, and advertising revenue. Understanding these motivations helps predict the direction of algorithm updates and adapt your strategy proactively rather than reactively.

Platforms consistently favor content that generates meaningful engagement, comments, shares, saves, and click-throughs, over passive consumption metrics like views or likes. Focus your content strategy on creating genuine interaction rather than optimizing for vanity metrics that don't translate to business results.

User satisfaction drives long-term algorithmic success because platforms need users to return consistently to maintain their advertising value. Content that educates, entertains, or inspires audiences to return regularly will always receive algorithmic support, regardless of specific technical changes.

Quality signals become increasingly important as platforms develop more sophisticated content evaluation methods. High production values, original content, expertise demonstration, and community building all send positive signals that transcend specific algorithm iterations.

Building Algorithm-Resistant Content

Content that maintains performance through algorithm changes shares common characteristics: it provides genuine value, encourages authentic engagement, demonstrates expertise, and builds community connections that extend beyond individual posts.

Educational content performs consistently because it serves clear user intent and generates engagement through questions, implementation shares, and result discussions. People actively seek educational content and engage meaningfully when they find valuable insights.

Community-building content that sparks discussions, asks thoughtful questions, and encourages user-generated responses creates the authentic engagement that all algorithms prioritize. Focus on content that generates conversations rather than passive consumption.

Algorithm Adaptation Strategies:

- **Diversification Approach**: Spread your content across multiple platforms and formats so algorithm changes on one platform don't devastate your entire strategy. Maintain an active presence on 3-4 platforms with different algorithmic approaches and audience demographics.
- **Engagement Velocity Focus**: Create content designed to generate immediate engagement through compelling hooks, controversial opinions, questions that encourage responses, and calls-to-action that prompt specific audience actions within the first hour of posting.
- **Content Format Flexibility**: Develop skills in multiple content formats (video, images, text, live streaming) so you can adapt quickly when platforms prioritize different content types through algorithm updates.
- **Audience Relationship Investment**: Build direct relationships with your most engaged followers through email lists, text messaging, personal interactions, and community building that doesn't depend on algorithmic distribution.
- **Original Content Creation**: Focus on creating unique, original content rather than following trends or repurposing others' ideas. Platforms increasingly favor original creators over content aggregators or trend followers.
- **Performance Monitoring Systems**: Establish tracking systems that quickly identify algorithm changes through performance metric shifts, allowing rapid strategy adjustments before significant audience or revenue impact occurs.

DIVERSIFYING PLATFORMS FOR STABILITY

Platform diversification protects your business from the risks of algorithm changes, platform policy updates, account restrictions, or entire platform decline. Building presence across multiple platforms creates stability while allowing you to capitalize on each platform's unique strengths and audience characteristics.

The most resilient social media businesses treat individual platforms as traffic sources rather than business foundations. This approach allows them to leverage platform benefits while maintaining business independence that survives platform-specific challenges or changes.

Effective diversification requires understanding each platform's unique culture, optimal content formats, and audience expectations rather than simply cross-posting identical content everywhere. Each platform should serve specific functions in your overall business strategy while contributing to common objectives.

Strategic Platform Selection

Choose platforms based on where your target audience spends time and which formats best showcase your expertise, rather than trying to maintain a presence on every available platform. Quality engagement on fewer platforms typically generates better business results than superficial presence everywhere.

Consider each platform's business model and monetization opportunities when making selection decisions. YouTube's ad revenue sharing differs significantly from Instagram's shopping features and TikTok's creator funds. Align platform selection with your revenue objectives and content strengths.

Platform demographics and user behavior should influence your content strategy and resource allocation. LinkedIn audiences expect professional content and longer-form insights, while TikTok users prefer quick entertainment and trending participation. Tailor your approach to match platform-specific expectations.

Emerging platform opportunities can provide significant advantages for early adopters who establish presence before competition intensifies. Monitor new platform developments and test promising opportunities while they still offer organic reach advantages.

Cross-Platform Content Strategy

Develop content themes and messages that work across multiple platforms while adapting format and presentation for each platform's unique requirements. This approach maximizes content value while respecting platform-specific user expectations.

Content adaptation involves more than resizing images or editing video lengths. Each platform requires different hooks, calls-to-action, hashtag strategies, and engagement approaches that feel native to that environment while serving your overall business objectives.

Timing strategies should consider optimal posting schedules for each platform based on your audience's behavior patterns. Your Instagram audience might be most active at different times than your LinkedIn connections, requiring platform-specific scheduling approaches.

Cross-promotion between platforms can drive audience growth and engagement when done strategically. Mention your other platforms naturally within content, share exclusive previews across different channels, and create content series that span multiple platforms while providing unique value on each.

Risk Management Through Diversification

Account backup strategies protect against platform-specific risks like account suspension, hacking, or policy violations that could eliminate years of audience building overnight. Maintain updated contact lists, content archives, and alternative communication methods for your most valuable community members.

Revenue stream diversification across platforms reduces dependence on any single platform's monetization features or policy changes.

Combine YouTube ad revenue, Instagram affiliate marketing, TikTok brand partnerships, and LinkedIn service sales to create stability through variety.

Audience relationship building should extend beyond platform-native features through email lists, text messaging, personal websites, and community platforms that you control independently of social media platforms. These owned channels provide stability when platform access becomes uncertain.

Content ownership and archiving ensures you maintain access to your best-performing content regardless of platform changes. Store original content files, successful caption templates, and performance data in systems you control rather than depending entirely on platform-provided analytics and storage.

BUILDING A LONG-TERM PASSIVE INCOME MACHINE

Creating sustainable passive income through social media requires building systems and assets that generate revenue with minimal ongoing input while remaining resilient to platform changes and market shifts. True passive income comes from owning valuable assets rather than depending on constant content creation and audience maintenance.

The transition from active content creation to passive income generation typically involves creating evergreen digital products, building automated sales systems, developing recurring revenue streams, and establishing investment vehicles that compound your social media earnings over time.

Long-term passive income sustainability requires balancing current revenue optimization with future asset development. Some of your time and resources must be invested in building systems and products that will generate income years from now, even if they don't contribute to immediate revenue goals.

Evergreen Asset Development

Digital assets like comprehensive courses, detailed guides, software tools, and template libraries can generate income for years after creation while requiring minimal ongoing maintenance. These assets leverage your expertise into scalable products that serve unlimited customers without proportional time investment.

Content libraries and searchable databases of your expertise become valuable assets that support multiple revenue streams through course content, consulting reference materials, book development, and speaking engagement preparation. Organize your best insights into accessible formats that provide ongoing value.

Intellectual property development through trademarked processes, copyrighted content, branded methodologies, and unique frameworks creates defensible business assets that competitors cannot replicate. These assets support premium pricing and partnership opportunities while building long-term business value.

Brand equity building through consistent quality, authentic relationship development, and expertise demonstration creates intangible assets that support all your monetization efforts. Strong personal brands command premium pricing and generate referral business that reduces marketing costs over time.

Automated Revenue Systems

Sales funnel automation moves prospects from social media awareness through email nurturing sequences to product purchases without requiring your direct involvement in each transaction. These systems work continuously to convert social media traffic into paying customers.

Customer onboarding automation delivers digital products, provides customer support, and encourages additional purchases through systematic follow-up sequences. Once established, these systems handle routine customer interactions while you focus on product development and strategy.

Subscription and membership systems generate recurring revenue through automated billing, content delivery, and community management. These systems provide predictable income streams while serving customers who want ongoing access to your expertise.

Affiliate income automation involves creating content that continues generating affiliate commissions long after publication through evergreen tutorials, product reviews, and educational content that includes affiliate links. This approach turns content creation into long-term income generation.

Investment and Wealth Building

Reinvestment strategies involve using social media profits to acquire income-generating assets like real estate, business investments, or financial instruments that provide returns independent of your ongoing content creation efforts.

Business expansion opportunities might include acquiring other creators' audiences, investing in complementary businesses, or developing new product lines that leverage your expertise and audience for additional revenue streams.

Team development and delegation allow your social media business to generate income while requiring less of your direct time investment. Well-trained team members can handle content creation, community management, and customer service while you focus on strategy and growth.

Exit strategy development involves building business systems and assets that have value to potential buyers, creating options for business sale or partnership arrangements that provide substantial returns on your social media business development efforts.

Legacy and Long-term Value Creation

Knowledge documentation ensures your expertise and successful strategies are preserved in formats that provide ongoing value through courses, books, consulting frameworks, and training programs that can outlast your active content creation period.

Mentorship and education programs allow you to share your knowledge while generating income through teaching others your successful methods. These programs create additional revenue streams while building your reputation as an industry expert.

Industry influence development through thought leadership, speaking engagements, advisory roles, and industry participation creates opportunities that extend beyond social media into broader professional and business contexts.

Succession planning for your social media business ensures that the assets and systems you've built can continue generating value through team members, business partners, or succession arrangements that preserve the long-term value of your work.

Action Steps for Chapter 12:

Action Item	Task Description	Expected Outcome
Algorithm Resilience Audit	Analyze your current content strategy for algorithm dependencies. Identify which tactics rely on specific platform features and develop alternative approaches that focus on fundamental value delivery and authentic engagement.	Strategy that maintains performance regardless of algorithm changes by focusing on timeless engagement principles

Action Item	Task Description	Expected Outcome
Platform Diversification Plan	Evaluate your current platform mix and develop a strategic plan for expanding to 2–3 additional platforms that serve your target audience. Create platform-specific content strategies that leverage each platform's unique strengths.	Reduced business risk through strategic presence across multiple platforms with tailored approaches for each
Passive Income Asset Creation	Identify and begin developing one major evergreen asset (comprehensive course, detailed guide, or automated system) that can generate revenue for years with minimal ongoing input. Plan the development timeline and resource requirements.	Foundation for long-term passive income through valuable digital assets that scale independently of your time investment
Automated Systems Implementation	Build or improve automated systems for customer acquisition, product delivery, and ongoing customer communication. Focus on systems that can operate effectively with minimal daily oversight while maintaining quality standards.	Business operations that generate income and serve customers automatically while you focus on strategic growth activities
Long-term Value Planning	Create a 5-year vision for your social media business that includes specific goals for passive income levels, business asset development, team growth, and potential exit or succession strategies.	Clear roadmap for building lasting wealth and business value beyond day-to-day content creation activities

Help Someone Take Their First Step Toward Online Success

Every act of sharing, whether it's knowledge, encouragement, or even a few kind words, has the power to ripple outward in ways we may never fully see.

Have you ever wondered how a single piece of advice, shared at just the right time, can spark someone else's breakthrough?

Imagine a reader who is standing where you once stood, curious about building an online business, unsure how to navigate the noise of **TikTok, Instagram, Facebook, YouTube, and SEO**, and maybe even doubting whether passive income is truly possible for them. Your experience with *Social Media Marketing: Crack the Code to Online Success With AI-Powered Content, SEO Mastery, and Proven Strategies to Monetize TikTok, Instagram, Facebook, YouTube, and Build Passive Income Streams* could be the guiding light that shows them it is.

That's why I'd be deeply grateful if you would leave an **honest review**. Your words don't just help me as the author, they help create a roadmap for the next reader searching for clarity, direction, and hope in their online journey.

By sharing your thoughts, you:

- **Empower future readers** to decide if this book is right for them.
- **Support the mission** of helping more people harness social media and AI to create freedom and income streams.
- **Encourage authenticity**, reminding others that success isn't one-size-fits-all.

Thank you, sincerely, for taking the time to share your reflections. And if you know someone else who dreams of turning social media into something more than scrolling and likes, I'd be honored if you passed this book along to them too.

Conclusion

You have just completed a comprehensive journey through the modern landscape of social media entrepreneurship. From understanding why social platforms have become the new economy to building automated systems that generate passive income, you now possess the complete roadmap that transforms social media from entertainment into a wealth-building machine.

HOW FAR YOU HAVE COME: AI, SEO, AND MONETIZATION MASTERY

When you started this book, social media might have felt overwhelming with constantly changing algorithms, endless content demands, and confusion about how to turn followers into income. Now you understand that success comes from systematic approaches rather than random tactics.

You have learned that AI is not replacing human creativity but amplifying it, allowing you to create more content faster while maintaining authenticity. You understand that SEO principles apply to social media, making your content discoverable not only today but for years to come. You have discovered that monetization happens naturally when you focus on serving your audience rather than selling to them.

The integration of these three elements, AI efficiency, SEO longevity, and authentic monetization, creates compound growth that accelerates over time. Each piece of optimized content you create builds authority. Every genuine relationship you develop creates referral opportunities. All the systems you implement reduce your workload while increasing your results.

You now recognize that social media success is not about viral moments or overnight transformation. It is about building sustainable systems that generate consistent results through valuable content, authentic relationships, and strategic business development. This understanding separates you from creators who chase trends and wonder why their growth does not last.

SHIFTING FROM CONSUMER TO CREATOR, FROM POSTING TO PROFITING

Your relationship with social media has fundamentally changed. Where you once saw distraction, you now see opportunity. Where you once consumed mindlessly, you now create strategically. Where you once spent money on these platforms, you now generate income through them.

This transformation goes beyond learning tactics. It represents a mindset shift from employee thinking to entrepreneur thinking. You no longer trade time for money; you build systems that generate income while you sleep. You do not just follow trends; you create content that starts conversations and influences decisions.

The skills you have developed extend far beyond social media. You understand human psychology, persuasive communication, business system development, and strategic thinking that applies to any entrepreneurial venture. Social media became your classroom for learning how to build sustainable, profitable businesses that serve others while creating freedom for yourself.

You have also learned that success does not require perfection or complete knowledge before starting. Every expert was once a beginner who decided to start before they felt ready. Your willingness to begin, learn from results, and adapt based on feedback will generate better outcomes than waiting for perfect conditions or complete confidence.

ENCOURAGEMENT: SMALL CONSISTENT ACTIONS CREATE BIG RESULTS

The strategies in this book work, but only when applied consistently over time. One piece of content will not build your empire. One AI-generated caption will not transform your business. One optimized post will not create passive income. But hundreds of valuable posts, systematic content creation, and persistent relationship building absolutely will.

Think compound interest, not immediate gratification. Every follower gained, every email subscriber added, and every piece of valuable content created adds to your business foundation. These actions may seem small daily, but they accumulate into significant results over months and years.

The creators who succeed long term are those who embrace the process rather than fixating on outcomes. They find satisfaction in helping their audiences, joy in creating valuable content, and excitement in building systems that serve others effectively. This process-focused approach makes the work sustainable and the results inevitable.

Remember that your unique combination of experiences, perspectives, and expertise cannot be replicated by anyone else. While others can learn similar strategies, they cannot duplicate your authentic voice, personal stories, or specific insights. This uniqueness is your competitive advantage, growing stronger as you develop your expertise and community.

FINAL CALL: YOUR TIME TO BUILD, GROW, AND THRIVE

The information age has created unprecedented opportunities for individuals to build global businesses from anywhere with an internet connection. Social media platforms provide free access to billions of potential customers. AI tools make content creation and business operations easier than ever. The infrastructure for online payments, customer service, and product delivery has never been more accessible.

Every day you delay starting is a day of opportunity lost. The audience who could benefit from your expertise is seeking solutions from other creators. The income you could be generating is flowing to competitors who decided to start before they felt ready. The business you could be building remains an unrealized possibility instead of a growing reality.

But starting does not mean perfection. Your first posts do not need to be masterpieces. Your initial products do not need to be comprehensive courses. Your early systems do not need to be complex operations. They just need to exist, serve others, and give you the experience that guides your next improvement.

The best time to plant a tree was twenty years ago. The second-best time is now. The best time to start building your social media empire was when these platforms launched. The second-best time is today, with all the tools, knowledge, and opportunities now available.

Your expertise has value. Your perspective matters. Your audience is waiting for someone exactly like you to help them achieve their goals. The only question that remains is whether you will step into this opportunity or keep observing from the sidelines.

The blueprint is in your hands. The tools are available. The audience is ready. Your time is now.

Start today. Start small. Start before you are ready. But start.

Your next chapter in online success begins with your next piece of content.

"The way to get started is to quit talking and begin doing."

– Walt Disney

PART II
Personal Branding

A COMPLETE SYSTEM TO BUILD AND MONETIZE A UNIQUE BRAND USING INFLUENCER STRATEGIES AND STEP-BY-STEP MARKETING TOOLS ON LINKEDIN, X, INSTAGRAM, YOUTUBE, FACEBOOK, AND TIKTOK

Introduction

The rules of business have fundamentally changed. Gone are the days when hiding behind corporate logos and company brands was enough to build a successful career or business. Today, people buy from people, not faceless corporations. They follow individuals who share their values, solve their problems, and inspire their aspirations.

THE SHIFT FROM CORPORATE IDENTITY TO PERSONAL IDENTITY IN BUSINESS

Traditional business operated on institutional trust. Companies spent decades building brand recognition through advertising, public relations, and corporate messaging. Employees were expected to be interchangeable parts in the corporate machine, their individual personalities absorbed into company culture and brand guidelines.

This model has become obsolete in the digital age. Consumers now expect transparency, authenticity, and personal connection from the brands they support. They want to know who stands behind the products and services they purchase. They seek relationships, not just transactions.

Consider how business decisions are made today. When choosing between two consultants with similar credentials, clients almost always select the one whose personal brand resonates with their values and communication style. When investors evaluate startups, they are betting on the founders' personal brands as much as the business concept. When companies hire executives, they consider not just professional qualifications but also how that person's personal brand aligns with and enhances their organizational goals.

The most successful professionals understand that their personal brand is their competitive advantage. It is what differentiates them in crowded markets, attracts opportunities, and commands premium pricing. Your personal brand becomes your business moat, something competitors cannot replicate because it is uniquely yours.

WHY PERSONAL BRANDING IS NON-NEGOTIABLE IN THE DIGITAL AGE

Digital platforms have democratized influence and entrepreneurship in ways never seen before. Anyone with expertise, passion, and strategic thinking can build a global audience and generate significant income. However, this same accessibility has created intense competition for attention and trust.

Without a strong personal brand, you become invisible in the digital noise. Your expertise remains hidden, your potential clients cannot find you, and opportunities pass to those who have mastered the art of personal branding. The choice is not whether to build a personal brand, but whether to build one intentionally or allow others to define it for you.

Search engines and social media algorithms favor individuals who consistently share valuable content and engage authentically with their communities. They reward personal brands that demonstrate expertise, build relationships, and create genuine value for their audiences. These platforms have become meritocracies where the most helpful, authentic, and consistent personal brands rise to the top.

Modern consumers conduct extensive research before making purchasing decisions. They examine your online presence, read your content, watch your videos, and evaluate your credibility before engaging with your services or products. Your personal brand serves as your twenty-four-hour sales representative, building trust and demonstrating value even when you are not directly involved in the conversation.

The global nature of digital business means your personal brand can attract opportunities from anywhere in the world. Geographic limitations no longer constrain professional growth when your expertise and personality can reach global audiences through strategic personal branding.

WHAT THIS BOOK PROVIDES: A COMPLETE SYSTEM FROM BUILDING TO MONETIZING

This book provides a comprehensive, step-by-step system for building and monetizing a personal brand across all major digital platforms. You will learn not just what to do, but how to do it efficiently and effectively, with specific strategies tailored to each platform's unique characteristics and audience expectations.

The system covers every aspect of personal branding: discovering your unique value proposition, crafting compelling messages, designing professional visual elements, building authority across multiple platforms, creating engaging content consistently, and converting your influence into sustainable income streams.

Each chapter builds upon the previous one, creating a cohesive strategy that compounds your efforts rather than scattering them across disconnected tactics. You will understand how content created for one platform can be adapted and repurposed for others, how different platforms serve different functions in your overall brand strategy, and how to create synergies that amplify your impact.

The strategies presented here are based on proven methods used by successful personal brands across various industries. They are practical, implementable, and designed to generate real results rather than just theoretical understanding. Every concept is accompanied by specific action steps that allow immediate implementation.

READER PROMISE: TRANSFORM VISIBILITY INTO INFLUENCE AND INFLUENCE INTO INCOME

By the end of this book, you will have a complete roadmap for transforming from hidden expert to visible authority to profitable influencer. This transformation does not happen overnight, but it follows predictable patterns that you can accelerate through strategic action and consistent execution.

You will understand how to identify and articulate your unique value in ways that resonate with your target audience. You will know how to create content that builds trust, demonstrates expertise, and naturally leads to business opportunities. You will have specific strategies for each major social platform and understand how to integrate them into a cohesive personal branding system.

Most importantly, you will have a clear path for monetizing your personal brand through multiple income streams including consulting, speaking, digital products, sponsorships, and partnership opportunities. You will understand how to price your expertise, attract premium clients, and build a sustainable business around your personal brand.

The journey from anonymity to influence requires commitment and consistent action, but the rewards extend far beyond financial success. A strong personal brand provides professional security, creative fulfillment, and the satisfaction of helping others while building something uniquely yours.

Your expertise has value. Your perspective matters. Your audience is waiting for someone exactly like you to help them achieve their goals. This book provides the systematic approach to connect your unique value with the people who need it most, creating mutual benefit that forms the foundation of sustainable business success.

THIRTEEN

Discovering Your Unique Brand Identity

Building a powerful personal brand begins with deep self-understanding. Most professionals skip this foundational step and jump directly to content creation or platform building, which explains why their brands feel generic and fail to attract engaged audiences. Your unique combination of experiences, knowledge, and perspectives cannot be replicated by anyone else, but you must first identify and articulate what makes you distinctive.

Effective self-assessment goes beyond listing job titles or educational credentials. It involves examining the intersection of your natural talents, developed skills, genuine interests, and life experiences that have shaped your worldview. This intersection is where your most authentic and compelling personal brand emerges.

The goal isn't to fabricate a persona or present a perfect image, but to identify the genuine aspects of your background and personality that provide unique value to others. Authenticity resonates because people instinctively recognize and trust genuine expertise paired with authentic personality.

Identifying Your Core Strengths

Your core strengths include both natural talents and developed capabilities that consistently produce superior results. These aren't just skills you possess, but areas where you naturally excel and others frequently seek your input or assistance.

Consider the compliments you receive regularly from colleagues, friends, or clients. What do people consistently praise about your work or approach? What tasks do others find challenging that come naturally to you? These patterns reveal strengths that you might take for granted but represent significant value to others.

Examine your professional history for recurring themes and consistent successes. Look beyond job titles to identify the specific contributions that generated the best outcomes in each role. What processes did you improve? What problems did you solve? What results did you achieve that others struggled to replicate?

Don't overlook strengths that seem unrelated to your current professional focus. Communication skills developed through teaching, organizational abilities honed through event planning, or analytical thinking refined through personal projects can all contribute to your unique professional value proposition.

Uncovering Your Genuine Passions

Passions provide the emotional fuel that sustains long-term personal brand building. Without genuine interest in your chosen focus area, you'll struggle to create consistent content, engage authentically with your audience, and maintain the persistence required for brand building success.

Identify topics that naturally capture your attention and energy. What subjects do you research voluntarily? What conversations energize rather than drain you? What problems do you solve in your spare time because you find them intellectually stimulating rather than burdensome?

Consider the intersection between your passions and market demand. The most sustainable personal brands operate at the convergence of what you love doing, what you're naturally good at, and what others value enough to pay for. This intersection ensures both personal satisfaction and business viability.

Don't assume your passions must align perfectly with traditional career paths. Many successful personal brands are built around unique combinations of interests that create distinctive niches. A fitness enthusiast with financial expertise might build a brand around financial wellness, while a technology professional with artistic interests might focus on creative applications of emerging technologies.

Cataloging Your Expertise and Experience

Your expertise extends far beyond formal credentials to include practical experience, problem-solving capabilities, and insights gained through real-world application of your knowledge. This practical wisdom often provides more value to your audience than theoretical knowledge alone.

Document your professional journey with attention to specific challenges you've navigated, solutions you've developed, and lessons you've learned through both successes and failures. These experiences provide the raw material for compelling content that resonates with others facing similar challenges.

Include expertise gained outside traditional professional settings. Volunteer work, personal projects, hobby pursuits, and life experiences all contribute to your unique knowledge base. Parents develop project management and conflict resolution skills, travelers gain cultural understanding and adaptability, and hobbyists often possess deep technical knowledge in specialized areas.

Consider the problems you solve naturally for friends, family, or colleagues. What questions do people ask you repeatedly? What situations do others find stressful that you handle easily? These patterns indicate areas of expertise that you might not formally recognize but that provide significant value to others.

DEFINING YOUR MISSION AND MESSAGE

Your mission and message form the foundation of everything you communicate through your personal brand. Without clarity about your fundamental purpose and core message, your content will lack focus and your audience will struggle to understand how you can help them achieve their goals.

A compelling mission statement answers why you do what you do and whom you serve, while your core message articulates the unique value you provide and the transformation you help others achieve. Together, these elements guide every decision about content creation, audience engagement, and business development.

The most powerful missions and messages emerge from the intersection of your personal experiences, professional expertise, and genuine desire to help others overcome challenges or achieve aspirations. This intersection ensures authenticity while creating clear value propositions that attract engaged audiences.

Crafting Your Mission Statement

Your mission statement should be specific enough to guide decision-making while broad enough to encompass your range of expertise and interests. It should clearly identify whom you serve and what transformation or outcome you help them achieve.

Effective mission statements focus on the impact you create rather than the services you provide. "I help overwhelmed entrepreneurs build systematic business operations" is more compelling than "I provide business consulting services" because it speaks directly to the audience's desired outcome.

Consider the deeper motivation behind your professional work. What injustice do you want to correct? What inefficiency do you want to eliminate? What possibility do you want to help others realize? This deeper purpose provides the emotional foundation that sustains long-term brand-building efforts.

Test your mission statement against your natural conversations about your work. When you're genuinely excited about helping someone, what do you say about what you do and why it matters? This natural language often provides better foundation material than formal business jargon.

Developing Your Core Message

Your core message distills your unique value proposition into clear, memorable language that immediately communicates why someone should pay attention to you rather than the many other experts in your space.

Strong core messages include specific outcomes, target audiences, and unique approaches or perspectives that differentiate you from others addressing similar problems. "I help technology leaders build inclusive teams through data-driven hiring practices" is more compelling than "I help companies with diversity and inclusion."

Your core message should be validated by your actual experience and results. If you claim to help businesses double their revenue, you should have concrete examples of achieving this outcome for clients. If you position yourself as an expert in work-life balance, your own life should demonstrate the principles you advocate.

Consider the language your ideal clients use to describe their challenges and aspirations. Your core message should speak their language and address their specific pain points using terminology they recognize and respond to emotionally.

Aligning Mission and Message with Market Needs

The most successful personal brands operate at the intersection of personal passion, unique expertise, and genuine market demand. All three elements must be present for sustainable brand-building success.

Research your target market to understand their most pressing challenges, unfulfilled needs, and desired outcomes. What problems keep them awake at night? What goals do they struggle to achieve? What solutions have they tried that failed to deliver promised results?

Validate demand for your unique approach through direct conversation with potential clients or audience members. Do they recognize the problems you solve? Do they value the outcomes you help achieve? Are they willing to invest time, attention, or money in solutions to these challenges?

Consider how your mission and message fit within the broader conversation in your industry or niche. What perspectives are missing? What approaches aren't being tried? What audiences aren't being served effectively by existing experts?

CRAFTING YOUR STORY FOR IMPACT AND RELATABILITY

Your personal story transforms abstract expertise into relatable human experience that creates emotional connections with your audience. People don't just buy products or services, they buy from individuals they trust, respect, and relate to on a personal level.

Effective brand storytelling doesn't require dramatic transformation narratives or extraordinary circumstances. The most compelling stories often emerge from ordinary challenges that your audience faces, showing how you navigated these difficulties and what you learned in the process.

Your story should highlight the journey from where you started to where you are now, with particular attention to the struggles, lessons, and insights that shaped your current expertise and perspective. This journey provides context for your knowledge while demonstrating empathy for others facing similar challenges.

Identifying Your Transformation Narrative

Every compelling personal brand story includes elements of transformation. How you moved from one state to another through effort, learning, and growth. This transformation doesn't need to be dramatic, but it should be genuine and relevant to your audience's aspirations.

Consider significant challenges you've overcome in your personal or professional life. What obstacles seemed insurmountable at the time? What fears did you face? What mistakes did you make? What breakthroughs changed your trajectory? These elements provide raw material for authentic storytelling.

Focus on transformations that relate to the problems you now help others solve. If you help businesses improve their operations, share how you learned to systematize your own work or organizational processes. If you teach productivity strategies, describe your journey from chaos to organized effectiveness.

The most powerful transformation narratives include details that make the story vivid and memorable. Rather than saying "I struggled with time management," describe the specific consequences of poor organization and the moment you realized something had to change.

Structuring Your Story for Maximum Impact

Effective brand stories follow predictable structures that create emotional engagement and memorable impact. The classic transformation arc moves from initial challenge through struggle and discovery to breakthrough and current state.

Begin with relatable circumstances that your audience recognizes—the problems, frustrations, or aspirations that many people in your target market experience. This creates an immediate connection and demonstrates your understanding of their situation.

Describe the journey of attempting to solve these problems, including failures, partial successes, and gradual learning that led to your current expertise. This journey phase shows persistence, learning ability, and the development of hard-won wisdom that provides value to others.

Conclude with the transformation you achieved and how this experience now enables you to help others navigate similar challenges more effectively. This conclusion positions you as a guide who has walked the path rather than a theorist who studied the map.

Making Your Story Relatable and Actionable

The best brand stories provide inspiration while offering practical insights that audience members can apply to their situations. Avoid stories that seem so unique or extraordinary that others cannot relate to your experience or imagine achieving similar outcomes.

Include specific strategies, mindset shifts, or practical actions that contributed to your transformation. What books did you read? What habits did you develop? What people influenced your thinking? What mistakes taught you valuable lessons? These details provide actionable guidance within your narrative.

Address common objections or limiting beliefs that might prevent others from taking action. If someone might think, "That worked for you, but it won't work for me," explain how your approach can be adapted to different circumstances or personality types.

Connect your story to universal themes and emotions that transcend specific industries or circumstances. Stories about overcoming fear, developing confidence, learning from failure, or discovering purpose resonate across diverse audiences while maintaining professional relevance.

Action Steps for Chapter 1:

Action Item	Task Description	Expected Outcome
Strengths and Expertise Inventory	Create a comprehensive list of your professional skills, natural talents, and unique experiences. Include feedback from others and recurring themes across your career journey.	Clear understanding of your unique value and competitive advantages
Mission Statement Development	Write a one-sentence mission statement that identifies who you serve and what transformation you help them achieve. Test it with trusted colleagues for clarity and impact.	Focused foundation that guides all brand decisions and communications
Core Message Crafting	Develop a 30-second elevator pitch that communicates your unique value proposition. Practice delivering it naturally in various professional situations.	Memorable, clear communication of what makes you distinctive and valuable
Transformation Story Documentation	Write your personal transformation story following the challenge-journey-breakthrough structure. Focus on experiences relevant to your target audience's aspirations.	Compelling narrative that builds trust and demonstrates your expertise through real experience
Market Validation Research	Conduct informal interviews with 5–10 people in your target market to validate that your mission and message address real needs and resonate with your intended audience.	Confirmation that your brand identity aligns with genuine market demand

FOURTEEN

Crafting a Magnetic Brand Message

Your elevator pitch serves as the cornerstone of all your brand communications. It's the distilled essence of who you are, what you do, and why it matters, delivered in a way that captures attention and invites further conversation. Unlike generic networking introductions that blend into forgettable noise, a magnetic elevator pitch creates immediate interest and positions you as someone worth knowing.

The most effective elevator pitches focus on the transformation you create rather than the services you provide. People don't buy consulting; they buy better business results. They don't hire coaches; they hire someone who helps them achieve their goals faster and more efficiently. Your pitch should immediately connect your expertise to outcomes your audience cares about.

A well-crafted elevator pitch adapts to different contexts while maintaining consistent core messaging. The version you use at industry conferences might emphasize different aspects than what you share at community networking events, but both should convey the same fundamental value proposition and leave listeners with a clear understanding of how you might help them.

The Problem-Solution-Outcome Formula

Structure your elevator pitch around the proven problem-solution-outcome formula that immediately establishes relevance and value. Begin by identifying a problem your audience faces, present your unique solution approach, and conclude with the specific outcome you help achieve.

"Most small business owners struggle with inconsistent marketing results because they're trying random tactics instead of following proven systems. I help them build marketing funnels that generate predictable leads and convert prospects into paying clients, typically increasing their revenue by 30-50% within six months."

This structure works because it follows natural human thinking patterns. People are drawn to those who understand their challenges, interested in solutions that make sense, and motivated by outcomes they want to achieve. The formula creates engagement at each step while building toward a compelling conclusion.

Avoid starting with your job title or credentials, which provide context but don't create emotional connection. "I'm a marketing consultant with an MBA and ten years of experience" tells people what you are, not what you do for them. Lead with the problem you solve, and credentials become supporting evidence rather than opening statements.

Quantifying Your Impact

Specific numbers and measurable outcomes transform vague promises into credible claims that capture attention and build confidence. Instead of saying you help people "improve their performance," specify that you "help executives increase their team's productivity by 25% while reducing overtime costs."

Use metrics that matter to your target audience. Financial professionals care about ROI and profit margins. Operations managers focus on efficiency and cost reduction. Marketing leaders want engagement rates and conversion percentages. HR executives value retention rates and employee satisfaction scores.

When you don't have specific client results to reference, use industry benchmarks or research data that supports your approach. "Studies show that companies with systematic onboarding processes see 60% better new hire retention," which provides a credible foundation for your onboarding expertise even without specific client case studies.

Be prepared to support your claims with brief examples or explanations when asked. Your elevator pitch creates interest, but follow-up questions require more detailed validation of your stated outcomes and approach.

Adapting Your Pitch for Different Audiences

Your core message remains consistent, but the emphasis and language should adapt to different audience contexts and interests. The problems that matter most to startup founders differ from those facing corporate executives, even when both might benefit from your expertise.

Research your audience before networking events or professional gatherings. What challenges does this industry face? What language do they use to describe their problems? What outcomes do they value most highly? This preparation allows you to customize your pitch for maximum relevance and impact.

Practice variations of your pitch that emphasize different aspects of your value. One version might focus on cost savings for budget-conscious audiences, while another emphasizes growth opportunities for expansion-minded leaders. The core transformation remains the same, but the framing adapts to different motivations.

Listen actively to how others introduce themselves and describe their challenges. This real-time market research helps you adjust your pitch to address the specific concerns and interests you hear expressed in that environment.

IDENTIFYING YOUR TARGET AUDIENCE AND IDEAL FOLLOWERS

Building a powerful personal brand requires absolute clarity about who you're trying to reach and influence. Many professionals try to appeal to everyone, but this results in messages that resonate with no one. Specific audiences with clearly defined needs, challenges, and aspirations form the foundation of the most successful personal brands.

Your target audience isn't just anyone who might potentially benefit from your expertise. It's the specific group of people you can help most effectively and who are most likely to engage with your content, share your message, and eventually become clients or customers.

Effective audience identification goes beyond demographic categories to include psychographic factors like values, beliefs, goals, frustrations, and communication preferences. Two people with identical job titles might have completely different attitudes toward professional development, risk tolerance, or decision-making processes.

Creating Detailed Audience Personas

Develop comprehensive profiles of your ideal audience members that go far beyond basic demographics. Include their professional responsibilities, personal goals, daily challenges, information consumption habits, and preferred communication styles.

Consider where your ideal followers spend their time online and offline. What publications do they read? What conferences do they attend? What social media platforms do they use most actively? What podcasts do they listen to during their commute? Understanding their media consumption patterns helps you position your content where they're most likely to discover it.

Document the language your audience uses to describe their challenges and aspirations. Do they talk about "work-life balance" or "professional fulfillment"? Do they focus on "increasing revenue" or "building sustainable businesses"? Using their language in your messaging creates immediate resonance and demonstrates understanding.

Include emotional and psychological factors in your audience profiles. What keeps them awake at night? What makes them feel most confident and capable? What fears hold them back from taking action? What aspirations motivate their professional decisions? These insights inform content that creates genuine connection.

Understanding Audience Pain Points and Desires

Your audience's pain points represent opportunities to provide immediate value through your content and expertise. The more accurately you can identify and articulate their specific challenges, the more your brand will resonate as the solution they've been seeking.

Distinguish between surface-level complaints and underlying root causes. Someone might complain about "not having enough time," but the real issue could be poor prioritization skills, lack of delegation capabilities, or inefficient systems and processes. Address the root causes, not just the symptoms.

Research the solutions your audience has already tried and why those approaches failed or fell short. This insight helps you position your approach as superior to existing alternatives while acknowledging their previous efforts and frustrations.

Identify aspirational outcomes that motivate your audience's decisions. What does success look like to them? What would their ideal professional life include? What recognition or achievements do they dream about? Understanding their positive motivations helps you frame your value proposition in inspiring rather than just problem-solving terms.

Analyzing Where Your Audience Congregates

Map the online and offline spaces where your ideal audience members gather to learn, network, and discuss industry topics. These locations become prime opportunities for building visibility and establishing your expertise within relevant communities.

Professional platforms like LinkedIn groups, industry forums, specialized communities, and association websites provide opportunities to engage directly with your target audience. Observe the conversations, questions, and concerns that generate the most engagement to understand what topics matter most.

Conference attendee lists, speaker lineups, and event agendas reveal where your audience invests their professional development time and budgets. These events provide networking opportunities while also indicating topics and themes that resonate with your target market.

Social media analytics from competitors and industry leaders show which platforms generate the most engagement from your target demographic. If your audience is most active on LinkedIn during business hours but engages with video content on YouTube during evenings and weekends, this information guides your content distribution strategy.

DEVELOPING A CONSISTENT VOICE AND STYLE

Your brand voice encompasses the personality, tone, and communication style that makes your content immediately recognizable and distinctly yours. Consistency in voice builds trust, creates expectation, and helps your audience develop a relationship with your brand that extends beyond individual pieces of content.

Developing an authentic brand voice requires understanding both your natural communication style and your audience's preferences and expectations. The goal isn't to create an artificial persona but to amplify the aspects of your authentic personality that resonate most effectively with the people you want to reach and influence.

Your voice should feel sustainable over time and across different content formats. If maintaining your chosen voice requires constant effort or feels unnatural, it will become inconsistent and ultimately ineffective. The best brand voices enhance rather than mask your authentic personality.

Defining Your Voice Characteristics

Identify three-to-five specific adjectives that capture the essence of how you want to communicate with your audience. These might include professional yet approachable, confident but humble, practical and results-focused, or innovative and forward-thinking. These characteristics guide every communication decision.

Consider the relationship dynamic you want to create with your audience. Do you want to be seen as the authoritative expert, the helpful peer, the challenging coach, or the encouraging mentor? This positioning influences your tone, language choices, and the level of vulnerability or authority you express.

Analyze your most successful communications, emails, presentations, or conversations that generated positive responses and strengthened relationships. What voice characteristics made those interactions effective? How can you consistently replicate those elements across all your brand communications?

Document specific language preferences and communication patterns that align with your brand voice. Do you use formal or conversational language? Do you prefer direct statements or questions that encourage engagement? Do you include humor, personal anecdotes, or stick to professional insights? These preferences create consistency across all your content.

Adapting Voice Across Platforms

While your core voice remains consistent, the expression of that voice adapts to different platform cultures and audience expectations. LinkedIn content might emphasize the professional and strategic aspects of your voice, while Instagram could highlight the personal and approachable elements.

Understand each platform's communication norms and user expectations. Twitter favors concise, often witty observations. LinkedIn rewards thoughtful, professional insights. YouTube allows for more casual, conversational communication. Instagram combines visual storytelling with authentic personality expression.

Maintain voice consistency through underlying values and perspectives rather than identical language and tone. Your core beliefs, expertise areas, and communication goals remain constant while the delivery method adapts to platform-specific requirements.

Test different voice expressions on various platforms to identify what generates the best engagement and feels most authentic to you. Some professionals discover they're naturally more effective with video content because their personality translates better through visual and audio cues than written text alone.

Creating Voice Guidelines and Standards

Document your voice characteristics, preferred language, and communication standards in a reference guide that ensures consistency across all your brand communications. This becomes especially important when you begin delegating content creation or collaborating with others on your brand development.

Include specific examples of how your voice handles different types of content—educational posts, promotional messages, responses to criticism, celebratory announcements, and personal shares. These examples provide templates for maintaining consistency across diverse communication scenarios.

Identify voice elements that conflict with your desired brand positioning. If you want to be seen as approachable, avoid overly technical jargon. If you're building authority, eliminate uncertain language like "I think" or "maybe." If you emphasize authenticity, avoid corporate speak and buzzwords.

Regularly review your content to ensure voice consistency and authenticity. Ask trusted colleagues or clients for feedback about how your brand voice comes across and whether it aligns with their experience of your personality and expertise in direct interactions.

Action Steps for Chapter 2:

Action Item	Task Description	Expected Outcome
Elevator Pitch Development	Create three versions of your elevator pitch using the problem-solution-outcome formula: a 30-second version, 60-second version, and 2-minute detailed version. Practice delivering each naturally in different scenarios.	Polished, compelling introduction that generates interest and positions you as a valuable expert
Audience Persona Creation	Develop detailed profiles of your top 3 ideal audience segments including demographics, psychographics, pain points, goals, and preferred communication channels. Interview actual people from each segment to validate your assumptions.	Deep understanding of who you're trying to reach and what messages will resonate most effectively
Voice Characteristics Definition	Identify 5 specific adjectives that describe your desired brand voice and create examples of how this voice handles different types of content (educational, promotional, personal, professional).	Clear, consistent communication style that feels authentic while appealing to your target audience

Action Item	Task Description	Expected Outcome
Message Testing Program	Test your elevator pitch and key messages with 10 people from your target audience. Gather feedback on clarity, relevance, and memorability to refine your core communications.	Validated messaging that resonates with real people and generates the responses you want
Platform Voice Adaptation	Adapt your core brand voice for each platform you plan to use, considering platform culture and audience expectations while maintaining consistency in values and perspective.	Platform-specific communication approaches that feel native while maintaining brand consistency

FIFTEEN

Designing Your Brand Look and Feel

Visual elements serve as the silent ambassadors of your personal brand, communicating professionalism, values, and personality before you speak a single word. In our image-saturated digital environment, strong visual branding isn't optional. It's essential for creating memorable first impressions and building recognition across platforms and interactions.

Your visual brand encompasses far more than just a logo. It includes your color palette, typography choices, photography style, graphic design elements, and the overall aesthetic that ties all your materials together. These elements work collectively to create a cohesive visual language that supports and reinforces your verbal messaging.

The most effective personal brand visuals strike a balance between professional credibility and authentic personality. Your visual choices should reflect both your expertise and your human warmth, creating an approachable yet authoritative presence that encourages connection while building trust.

Color Psychology and Palette Selection

Colors trigger emotional responses and psychological associations that operate below conscious awareness but significantly influence how others perceive your brand. Understanding color psychology allows you to make strategic choices that support your desired brand positioning and audience response.

Blue conveys trust, stability, and professionalism, making it popular among financial advisors, consultants, and corporate professionals. Red suggests energy, passion, and urgency, working well for motivational speakers, fitness experts, and sales professionals. Green implies growth, harmony, and prosperity, appealing to environmental consultants, wellness coaches, and financial planners.

However, avoid choosing colors based solely on industry conventions. Your color palette should align with your authentic personality and differentiate you from competitors who default to expected choices. A financial advisor who uses warm oranges and teals can stand out while still maintaining professionalism through thoughtful application.

Develop a primary palette of 2-3 colors that work harmoniously together and can be applied consistently across all your brand materials. Include neutral colors (whites, grays, or beiges) that provide breathing space and ensure your primary colors have maximum impact when used strategically.

Typography and Font Choices

Typography choices communicate personality and professionalism as powerfully as color selections. Fonts carry emotional weight and cultural associations that influence how others perceive your credibility and expertise before they read your content.

Serif fonts (like Times New Roman or Georgia) suggest tradition, reliability, and academic credibility, making them appropriate for lawyers, professors, or consultants who want to emphasize their scholarly expertise. Sans-serif fonts (like Arial or Helvetica) appear modern, clean, and approachable, working well for technology professionals, designers, or coaches who want to seem current and accessible.

Script fonts can add personality and warmth but should be used sparingly and only when legibility isn't compromised. They work best for creative professionals like photographers, designers, or lifestyle coaches who want to emphasize their artistic sensibility.

Establish a typography hierarchy that includes fonts for headlines, body text, and accent elements. Consistency in font choices across all your materials creates visual cohesion while ensuring your content remains readable across different platforms and devices.

Essential Visual Branding Elements:

- **Professional Headshots**: Invest in high-quality photography that captures your personality while maintaining professional standards. Your headshot should be consistent across all platforms and updated regularly to reflect your current appearance and brand evolution.
- **Logo Design**: Create a simple, memorable mark that works effectively at various sizes and across different backgrounds. Your logo should be scalable from business card size to billboard dimensions while maintaining clarity and impact.
- **Brand Pattern Library**: Develop consistent patterns, shapes, or design elements that create visual recognition and tie your materials together. These might include specific border styles, background textures, or graphic elements that appear across your content.
- **Image Filter/Style**: Establish consistent photo editing approaches that create visual cohesion across your content. This might include specific color adjustments, contrast levels, or filter applications that make your images immediately recognizable.

- **Template Systems**: Design templates for common content types (social media posts, presentations, documents) that ensure consistency while allowing efficient content creation. Templates save time while maintaining professional visual standards.
- **Brand Guidelines Document**: Create a reference guide that documents all your visual choices, including color codes, font specifications, logo usage rules, and application examples. This ensures consistency as your brand grows and evolves.

CREATING A PROFESSIONAL ONLINE PRESENCE

Your online presence serves as your digital headquarters, often providing the first comprehensive impression potential clients, employers, or collaborators have of your expertise and professionalism. Every element of your digital presence should work together to build credibility, demonstrate value, and encourage the next step in your professional relationship.

Creating a professional online presence extends beyond having profiles on social platforms. It includes your website, email signature, online portfolios, professional directories, and any other digital touchpoints where people might encounter your brand.

The goal isn't perfection but consistency and professionalism that accurately represents your capabilities while making it easy for others to understand how you might help them achieve their goals.

Website Development and Optimization

Your professional website serves as the central hub of your online presence, providing comprehensive information about your expertise while giving you complete control over the user experience and messaging.

Effective personal brand websites include clear value propositions, detailed service descriptions, client testimonials, comprehensive about pages, and obvious contact information. Visitors should understand within seconds what you do, whom you serve, and how to engage your services.

Include a professional bio that tells your story while emphasizing the experiences and credentials most relevant to your target audience. Your bio should balance professional accomplishments with personality elements that help visitors connect with you as a person.

Showcase your expertise through case studies, client results, media mentions, speaking engagements, and content samples. Social proof builds credibility while demonstrating the real-world impact of your work.

Optimize your website for search engines by including relevant keywords in your content, creating valuable resources that attract inbound links, and maintaining technically sound site performance. Many potential clients will discover you through search engines rather than social media or referrals.

Social Media Profile Optimization

Each social media platform requires platform-specific optimization that considers the unique culture and expectations of its user base while maintaining consistency with your overall brand message.

LinkedIn profiles should emphasize professional achievements, specific skills, and clear value propositions that help connection requests and business inquiries. Include keywords that your ideal clients might use when searching for your type of expertise.

Instagram profiles need compelling visuals, clear bio descriptions, and strategic use of highlights to showcase different aspects of your expertise. The visual nature of the platform requires more attention to aesthetic consistency and personality expression.

Twitter profiles should capture your expertise and personality in very limited space while demonstrating your engagement with industry conversations. Your tweet history becomes part of your profile, so maintain consistency with your desired brand positioning.

Ensure all your social profiles include professional headshots, consistent bio information, and clear contact information that makes it easy for interested parties to connect with you outside the platform.

Content Portfolio Development

Develop a comprehensive portfolio of your best work that demonstrates the range and quality of your expertise. This portfolio should be easily accessible and regularly updated with your most impressive recent accomplishments.

Include various content types that showcase different aspects of your capabilities: written articles that demonstrate thought leadership, video content that shows your communication skills, project results that prove your effectiveness, and testimonials that validate your client relationships.

Organize your portfolio strategically to guide visitors toward the type of engagement you most want to encourage. If you're seeking speaking opportunities, emphasize video content and previous speaking engagements. If you're building consulting business, highlight case studies and client results.

Make your portfolio easily shareable and accessible across all your professional platforms. Potential clients should be able to quickly access examples of your work regardless of where they first encounter your brand.

BRAND CONSISTENCY ACROSS PLATFORMS

Consistency across all platforms and touchpoints creates professional credibility while building recognition that compounds over time. When your visual elements, messaging, and personality remain consistent, people begin to recognize your content and associate it with quality and reliability.

However, consistency doesn't mean identical. Each platform has unique requirements and user expectations that require thoughtful adaptation while maintaining core brand elements that tie everything together.

The key is identifying which elements remain constant (your core colors, fonts, logo, and key messages) and which elements adapt to platform-specific requirements (image dimensions, content formats, and communication tone).

Platform-Specific Adaptation Strategies

LinkedIn requires a more formal, professional presentation that emphasizes business value and industry expertise. Your content should be comprehensive, well-researched, and focused on professional development or industry insights.

Instagram needs visually appealing content that showcases personality alongside expertise. The platform culture encourages more personal sharing and behind-the-scenes content that humanizes your professional brand.

Twitter favors concise, timely commentary on industry trends and current events. Your content should demonstrate thought leadership while engaging in broader conversations relevant to your expertise.

YouTube allows for comprehensive education and personality expression through longer-form content. Your videos should demonstrate teaching ability while showcasing both expertise and communication skills.

Facebook serves various functions depending on whether you use personal profiles, business pages, or groups. The platform works well for community building and more casual professional networking.

Visual Consistency Systems

Develop templates and style guides that ensure visual consistency while accommodating platform-specific requirements. Your Instagram posts might use different dimensions than your LinkedIn articles, but they should share color palettes, fonts, and design elements that create immediate recognition.

Create branded templates for common content types: quote graphics, educational carousels, video thumbnails, and presentation slides. These templates save time while ensuring professional appearance and brand consistency.

Maintain photo editing consistency through preset filters or editing approaches that create a cohesive visual style across all your imagery. Your photos should look like they belong to the same brand regardless of where they appear.

Use consistent graphic elements, color schemes, and typography across all platforms while adapting layouts and formats to meet platform-specific requirements and user expectations.

Message Consistency Framework

Your core value proposition, key messages, and fundamental positioning should remain consistent across all platforms while the presentation and emphasis adapt to different audience segments and platform cultures.

Develop a master message document that includes your elevator pitch, key talking points, value propositions, and core content themes. This document ensures consistency while providing flexibility for platform-specific adaptation.

Create content calendars that coordinate messaging across platforms, ensuring your audience receives consistent value regardless of where they follow your content. Avoid contradictory messages or positioning that might confuse your audience.

Regularly audit your brand presence across all platforms to identify inconsistencies and opportunities for improvement. Your brand should feel cohesive and professional regardless of where people encounter it.

Action Steps for Chapter 3:

Action Item	Task Description	Expected Outcome
Visual Brand Elements Creation	Develop your color palette (2–3 primary colors plus neutrals), select brand fonts, and create or commission a professional logo. Document all specifications in a brand guidelines reference.	Complete visual identity system that can be applied consistently across all brand materials
Professional Photography Session	Invest in professional headshots and brand photography that captures your personality while maintaining professional standards. Get multiple options for different contexts and platforms.	High-quality visual assets that enhance credibility and create strong first impressions
Website Development	Build or update your professional website with clear value propositions, comprehensive about page, service descriptions, testimonials, and contact information. Optimize for search engines and mobile devices.	Professional online headquarters that effectively represents your expertise and converts visitors to clients
Social Media Optimization	Update all your social media profiles with consistent branding elements, professional photos, optimized bio descriptions, and relevant contact information. Ensure platform-specific optimization while maintaining brand consistency.	Cohesive online presence that builds recognition and makes it easy for prospects to connect with you
Brand Guidelines Documentation	Create a comprehensive brand guide that documents all visual elements, messaging frameworks, and application examples. Include specific instructions for maintaining consistency across different contexts.	Reference system that ensures brand consistency as you create content and expand your presence

SIXTEEN

Building Authority on LinkedIn

LinkedIn serves as the world's largest professional networking platform and represents the most important digital real estate for building business credibility and attracting high-value opportunities. Your LinkedIn profile functions as your professional headquarters, often providing the first comprehensive impression potential clients, employers, or collaborators have of your expertise and accomplishments.

Optimizing your LinkedIn profile requires strategic thinking beyond simply listing your job history. Every section should work together to tell a compelling story about your unique value while making it easy for your ideal audience to understand how you can help them achieve their goals.

The most effective LinkedIn profiles balance professional achievement with authentic personality, creating connections that feel both credible and approachable. Your profile should demonstrate expertise while showing the human side of your professional brand.

Headline Optimization for Maximum Impact

Your LinkedIn headline appears in search results, connection requests, and comment threads, making it one of your most valuable branding real estate opportunities. Rather than defaulting to your job title, craft a headline that communicates your value proposition and target audience.

"Marketing Director at TechCorp" tells people what you are but not what you do for others. "I help B2B SaaS companies increase qualified leads by 40% through strategic content marketing" immediately communicates value and attracts the right audience.

Include relevant keywords that your ideal connections might use when searching for expertise like yours. If you help companies with digital transformation, include terms like "digital strategy," "business transformation," or "technology implementation" that prospects might search for.

Keep your headline current and relevant to your goals. If you're transitioning careers, pivoting industries, or launching new services, your headline should reflect your desired positioning rather than just your current role.

About Section Storytelling

Your About section provides space for comprehensive storytelling that builds connection while demonstrating expertise. This section should balance professional accomplishments with personality elements that help visitors understand both your capabilities and your character.

Start with a compelling hook that immediately captures attention and speaks directly to your target audience's needs or aspirations. "After watching too many great ideas fail due to poor execution, I became obsessed with helping entrepreneurs build systematic business operations" creates immediate interest and positions expertise.

Include specific results and accomplishments that demonstrate your effectiveness. Quantify your impact wherever possible: "increased client retention rates by 35%," "generated $2.3M in new revenue," or "reduced operational costs by 22%." Numbers provide credible evidence of your capabilities.

Share your professional philosophy and approach in language that resonates with your target audience. What principles guide your work? What methods do you use that differ from others in your field? This insight helps prospects understand your unique approach and values alignment.

Conclude with a clear call-to-action that invites connection and engagement. "Let's connect if you're interested in discussing how systematic operations can accelerate your business growth" provides specific direction for interested visitors.

Experience Section Strategic Positioning

Present your work experience in ways that highlight transferable skills and relevant accomplishments rather than just listing job duties. Each role should demonstrate progression, learning, and results that support your current professional positioning.

Focus on achievements and outcomes rather than responsibilities and tasks. Instead of "managed marketing campaigns," write "developed content marketing strategy that increased lead generation by 65% while reducing cost-per-lead by 30%."

Include volunteer work, board positions, and significant projects that demonstrate relevant expertise and community involvement. These experiences often provide differentiation and show commitment to your professional field.

Use consistent formatting and professional language throughout your experience descriptions. Maintain active voice and quantify results wherever possible to create compelling evidence of your effectiveness.

Strategic LinkedIn Profile Optimization Tactics:

1. **Skills Section Curation**: Select and prioritize the 10-15 skills most relevant to your target audience and professional goals. Regularly update this section to reflect your current focus areas and encourage endorsements from colleagues and clients who can validate your expertise.
2. **Recommendations Strategy**: Proactively request recommendations from clients, colleagues, and supervisors who can speak specifically to your results and character. Offer to write recommendations for others first, which often generates reciprocal recommendations that enhance your credibility.
3. **Featured Section Utilization**: Use the Featured section to showcase your best work including articles, videos, presentations, or external content that demonstrates your expertise. This section provides visual interest while highlighting your most impressive professional accomplishments.
4. **Contact Information Optimization**: Make it easy for interested prospects to reach you by including multiple contact methods and links to your other professional platforms. Consider using a scheduling link if you regularly take discovery calls or consultations.

PUBLISHING CONTENT THAT POSITIONS YOU AS A THOUGHT LEADER

LinkedIn's publishing platform and native content features provide powerful opportunities to demonstrate expertise, share insights, and build authority within your professional community. Consistent, valuable content creation establishes you as a go-to resource while attracting the attention of potential clients, collaborators, and opportunities.

Thought leadership content differs from promotional material by focusing on industry insights, professional development, and valuable perspectives that help others succeed. The best LinkedIn content provides immediate value while subtly demonstrating the author's expertise and unique approach.

Content that performs well on LinkedIn balances professional value with authentic personality, creating posts that feel both informative and relatable. Your content should demonstrate not just what you know, but how you think and why others should care about your perspective.

Content Types That Build Authority

Industry analysis and trend commentary position you as someone who understands the broader context of your field and can help others navigate change. Share your perspective on industry developments, regulatory changes, market shifts, or emerging opportunities that affect your audience.

Case studies and client success stories (with appropriate permissions) demonstrate your ability to generate real results for others. These stories provide social proof while showing your methodology and approach in action.

Behind-the-scenes insights about your work process, decision-making approaches, or business operations humanize your expertise while providing valuable learning for others facing similar challenges.

Professional development content including lessons learned, skill-building advice, and career guidance appeals to LinkedIn's professional audience while demonstrating your experience and wisdom.

Content Structure for Maximum Engagement

Start your posts with compelling hooks that immediately capture attention and promise value. Questions, surprising statistics, bold statements, or personal anecdotes can draw readers into your content.

Use short paragraphs and line breaks to create scannable content that works well on mobile devices. LinkedIn users often browse during brief moments throughout their workday, so make your content easy to consume quickly.

Include specific examples, data, or stories that support your main points. Abstract advice becomes more valuable when illustrated with concrete applications that readers can relate to their own situations.

End posts with questions or calls-to-action that encourage engagement. The LinkedIn algorithm favors content that generates comments and discussion, so explicitly invite reader interaction.

Content Calendar and Consistency Strategy

Develop a content calendar that balances different types of posts while maintaining regular publishing frequency. Consistency builds audience expectations and algorithmic favor, but quality should never be sacrificed for quantity.

Plan content themes around your expertise areas, industry events, seasonal topics, and current news that affects your professional community. This strategic approach ensures relevance while demonstrating your thought leadership across various aspects of your field.

Repurpose your best content across different formats—turn a successful post into a longer article, create an infographic from popular insights, or develop video content around your most engaging written posts.

Monitor your content performance to identify topics, formats, and posting times that generate the best engagement with your specific audience. Use these insights to optimize future content strategies.

LEVERAGING CONNECTIONS, NETWORKING, AND LEAD GENERATION

LinkedIn's networking capabilities extend far beyond collecting connections to building meaningful professional relationships that generate business opportunities, collaborations, and career advancement. Strategic networking on LinkedIn requires systematic approaches that provide value to others while building your own professional network and opportunities.

The platform's advanced search capabilities, industry groups, and messaging features provide powerful tools for identifying, connecting with, and building relationships with your ideal professional contacts. However, success requires authentic relationship building rather than transactional networking approaches.

Effective LinkedIn networking focuses on providing value, building genuine relationships, and establishing yourself as a valuable connection worthy of referrals, collaborations, and business opportunities.

Strategic Connection Building

Identify specific types of professionals who could benefit from your expertise or provide valuable opportunities for collaboration, referrals, or business development. This might include potential clients, referral partners, industry peers, or thought leaders in your field.

Use LinkedIn's advanced search features to find professionals based on specific criteria including industry, company size, geographic location, seniority level, and mutual connections. These targeted searches help you identify high-value connection opportunities.

Personalize every connection request with specific reasons for connecting and potential value you could provide. Generic connection requests get ignored, while thoughtful messages that demonstrate genuine interest and potential mutual benefit generate positive responses.

Follow up on accepted connections with valuable content, helpful introductions, or relevant opportunities rather than immediate sales pitches. Relationship building takes time and consistent value delivery before business opportunities naturally emerge.

LinkedIn Groups and Community Engagement

Participate actively in LinkedIn groups where your ideal clients and professional peers gather to discuss industry topics and challenges. Group participation positions you as an expert while building relationships with potential clients and referral sources.

Share valuable insights and answer questions in group discussions rather than promoting your services directly. Helpful contributions build your reputation within the community and encourage private messages and connection requests from interested members.

Start meaningful discussions by posting thoughtful questions or sharing industry insights that generate engagement and position you as a thought leader within the group community.

Use group membership lists to identify potential connections who are already engaged in your professional community and likely to be interested in building relationships with relevant experts.

Lead Generation and Business Development

Use LinkedIn Sales Navigator or advanced search features to identify potential clients based on specific criteria relevant to your services. Research their challenges, recent company developments, and mutual connections that could facilitate introductions.

Engage with prospects' content before reaching out directly. Like, comment meaningfully on, and share their posts to build familiarity and demonstrate genuine interest in their work and perspectives.

Craft personalized outreach messages that focus on their challenges and how your expertise might help rather than immediately pitching your services. The goal is to start conversations that could develop into business relationships over time.

Track your networking activities and follow up systematically with connections who have expressed interest in your services. Use LinkedIn's note-taking features to remember conversation details and maintain personalized communication.

Action Steps for Chapter 4:

Action Item	Task Description	Expected Outcome
LinkedIn Profile Optimization	Complete a comprehensive profile update including compelling headline, detailed About section, optimized experience descriptions, and strategic skills selection. Include professional photos and contact information.	Professional LinkedIn presence that effectively communicates your value and attracts your ideal audience
Content Publishing Strategy	Develop a content calendar for LinkedIn posting including 20 initial post ideas across different content types (industry insights, case studies, professional development). Commit to posting 3-5 times per week.	Consistent thought leadership presence that builds authority and engagement with your target audience
Strategic Connection Campaign	Identify 100 potential connections in your target market and send personalized connection requests. Follow up with accepted connections using value-first approaches rather than sales pitches.	Expanded professional network of relevant contacts who could become clients, collaborators, or referral sources

Action Item	Task Description	Expected Outcome
Group Engagement Program	Join 5-10 LinkedIn groups where your ideal clients gather and commit to participating actively by answering questions, sharing insights, and starting valuable discussions.	Increased visibility and credibility within your professional community leading to connection requests and business opportunities
Lead Generation System	Set up a systematic approach for identifying and reaching out to potential clients on LinkedIn. Create templates for outreach messages and establish a follow-up schedule for nurturing relationships.	Predictable pipeline of qualified prospects generated through strategic LinkedIn networking and relationship building

SEVENTEEN

Becoming a Voice on X (Twitter)

X (formerly Twitter) represents the global town square for real-time professional discourse, breaking news, and thought leadership across every industry and discipline. The platform's character limit forces concise, impactful communication that cuts through noise and delivers immediate value to busy professionals seeking quick insights and timely perspectives.

Building influence on X requires mastering the art of distilling complex ideas into memorable, shareable insights that demonstrate expertise while sparking meaningful conversations. The most successful thought leaders on the platform combine deep knowledge with accessible communication, making sophisticated concepts understandable and actionable for broad audiences.

Unlike other platforms that favor polished content, X rewards authenticity, timeliness, and genuine engagement with current events and industry developments. Your influence grows through consistent valuable contributions to ongoing conversations rather than broadcasting promotional messages.

The Art of Concise Thought Leadership

Effective X thought leadership transforms complex professional insights into clear, actionable statements that busy professionals can quickly understand and apply. This requires understanding your core message and expressing it with precision and impact.

Start with comprehensive knowledge and distill it down to essential elements that provide immediate value. If you understand a complex business process, identify the three most critical success factors. If you've learned from a major failure, extract the key lesson that others can apply to avoid similar mistakes.

Use specific examples and concrete language rather than abstract concepts. "Successful product launches require clear success metrics, realistic timelines, and cross-functional alignment" is more valuable than "Product management involves strategic planning and execution excellence."

Test different ways of expressing the same insights to identify language and framing that generates the most engagement and resonates most strongly with your target audience. The same principle might be expressed in multiple ways that appeal to different professional perspectives.

Voice Development for Professional Impact

Your X voice should reflect your authentic professional personality while adapting to the platform's conversational, immediate culture. The best professional voices on X combine expertise with approachability, creating content that feels both authoritative and relatable.

Develop signature phrases, perspectives, or frameworks that become associated with your brand. This might be a particular way of analyzing industry trends, a consistent methodology you apply to problems, or a unique perspective that differentiates your insights from others in your field.

Balance professional insights with personality elements that make your content memorable and shareable. Humor, personal anecdotes, and candid observations can enhance rather than detract from your professional credibility when used appropriately.

Maintain consistency in your core values and perspectives while adapting your communication style to different topics and audiences. Your fundamental expertise and worldview should remain recognizable across all your content.

Timing and Relevance Strategies

X rewards timeliness and relevance more than other platforms, making it essential to stay current with industry developments, news events, and trending topics that affect your professional community.

Monitor industry news, company announcements, and economic developments that create opportunities for valuable commentary from your expertise perspective. Quick, insightful responses to breaking news can generate significant visibility and engagement.

Participate in industry conversations while they're happening rather than commenting after interest has moved elsewhere. Set up alerts for keywords, company names, and industry terms that might generate opportunities for valuable contribution.

Connect current events to timeless professional principles and insights. When news breaks in your industry, help others understand the broader implications, lessons learned, or strategic considerations that extend beyond the immediate story.

THREADS AND VIRAL STRATEGIES

X threads allow for longer-form storytelling and comprehensive insight sharing while maintaining the platform's conversational feel and engaging format. Well-crafted threads can go viral, dramatically expanding your reach and establishing your expertise with new audiences who discover your content through shares and engagement.

Thread creation requires understanding both storytelling principles and X's unique culture and algorithm preferences. The most successful threads combine valuable information with an engaging narrative structure that keeps readers clicking through to the conclusion.

Viral content on X typically provides either immediate practical value or strong emotional resonance, often combining both elements to create content that people feel compelled to share with their networks.

Thread Structure and Storytelling

Begin threads with compelling hooks that promise specific value or intriguing insights. Your opening tweet should clearly communicate what readers will gain by investing time in your full thread.

Structure threads like engaging stories with clear beginning, middle, and end sections. Even educational or analytical threads benefit from narrative progression that maintains interest throughout the entire sequence.

Use numbered tweets to help readers follow the logical progression and know how much content remains. This structure also makes threads easier to bookmark and reference later.

Include visual elements, data, or examples that break up text-heavy content and provide additional value. Screenshots, charts, or infographics can significantly increase engagement and shareability.

Conclude threads with clear takeaways and calls to action that encourage engagement, follows, or further discussion. Strong endings often determine whether readers share content with their networks.

Viral Content Principles

Viral X content typically addresses universal professional challenges, provides actionable insights, or offers unique perspectives on widely discussed topics. The key is finding fresh angles on subjects that already have audience interest and engagement.

Educational threads that teach valuable skills or share hard-won lessons tend to perform well because they provide immediate practical value that readers want to save and share with colleagues.

Contrarian takes on popular opinions can generate significant engagement and discussion, but ensure your perspective is well-reasoned and adds genuine value to the conversation rather than being controversial for attention alone.

Personal stories that illustrate broader professional principles often resonate strongly because they combine relatability with actionable insights. Share failures, learning experiences, and breakthrough moments that others can learn from.

Engagement and Community Building

Respond actively to comments and questions on your threads to encourage continued discussion and signal to the algorithm that your content is generating valuable engagement.

Engage with others who share or comment on your threads by visiting their profiles, following valuable contributors, and participating in their content discussions. This reciprocal engagement builds relationships and expands your network.

Use thread success to identify topics and formats that resonate most strongly with your audience, then create additional content around similar themes while avoiding exact repetition.

Track which threads generate the most profile visits, follows, and business inquiries to understand what content best serves your professional objectives beyond just engagement metrics.

ENGAGING IN INDUSTRY CONVERSATIONS FOR VISIBILITY

X's real-time nature makes it the primary platform for industry discussions, breaking news commentary, and professional debate across virtually every field. Active participation in these conversations positions you as an engaged expert while building relationships with other industry voices and potential clients or collaborators.

Successful industry engagement requires balance between contributing valuable insights and avoiding controversial topics that could damage your professional reputation. The goal is to become known as a thoughtful contributor whose perspectives add value to ongoing discussions.

Strategic conversation participation builds your network, increases your visibility, and establishes you as someone worth following for industry insights and professional development.

Identifying High-Value Conversations

Monitor hashtags, trending topics, and ongoing discussions relevant to your industry to identify opportunities for valuable contribution. Use X's search functions and lists to track conversations among industry leaders and potential clients.

Follow industry publications, thought leaders, and companies whose content regularly generates meaningful professional discussions. These accounts often serve as conversation starters that create opportunities for valuable engagement.

Set up keyword alerts for terms related to your expertise, industry developments, or client challenges that might create opportunities for helpful responses or insightful commentary.

Participate in recurring industry conversations like weekly hashtag discussions, live-tweeting industry events, or commenting on quarterly earnings reports that affect your professional community.

Strategic Engagement Approaches

Add substantive value to conversations rather than simply agreeing or sharing obvious observations. Provide additional context, alternative perspectives, or practical applications that enhance the original discussion.

Ask thoughtful questions that advance conversations and demonstrate your expertise while encouraging others to share their perspectives and experiences.

Share relevant resources, tools, or insights that help others in the conversation achieve better outcomes or understand complex topics more clearly.

Acknowledge others' contributions and build on their insights rather than contradicting or dismissing different viewpoints, which builds relationships while maintaining professional credibility.

Building Professional Relationships

Use industry conversations as opportunities to identify and connect with other professionals whose work and perspectives you respect and who might become valuable network connections.

Follow up on meaningful conversation exchanges with direct messages or connection requests that reference your interaction and suggest potential mutual value.

Offer to help others in conversations when you have relevant expertise or resources that could benefit them, building goodwill and professional relationships.

Share and amplify others' valuable content with thoughtful commentary that adds to their insights while building reciprocal relationships with industry peers.

Monitoring and Managing Your Professional Reputation

Regularly review your X activity to ensure your professional reputation remains positive and your contributions consistently add value to industry discussions.

Avoid engaging in heated debates or controversial topics that could alienate potential clients or damage your professional reputation, even when you have strong opinions about industry issues.

Focus on being helpful, insightful, and professional in all your interactions, remembering that potential clients and collaborators may review your X history when evaluating whether to work with you.

Use X's analytics to track which types of engagement generate the most positive responses and business inquiries, focusing your conversation participation on the most professionally valuable activities.

Action Steps for Chapter 5:

Action Item	Task Description	Expected Outcome
X Profile and Voice Development	Optimize your X profile with professional photo, compelling bio, and relevant links. Develop your unique voice by posting 10–15 tweets that demonstrate your expertise and personality, testing different tones and topics.	Established X presence with clear professional positioning and authentic voice that resonates with your target audience
Thread Content Creation	Plan and publish 5 high-value threads covering different aspects of your expertise. Use storytelling structure with compelling hooks, valuable insights, and clear calls-to-action. Monitor performance to identify successful formats.	Demonstrated thought leadership through comprehensive content that showcases expertise while building audience engagement

Action Item	Task Description	Expected Outcome
Industry Conversation Mapping	Identify 20 industry leaders, publications, and hashtags to follow for conversation opportunities. Set up keyword alerts and create lists to monitor relevant discussions in your professional area.	Systematic approach to discovering and participating in high-value professional conversations that build visibility and credibility
Daily Engagement Strategy	Establish a daily routine of posting valuable content, responding to industry discussions, and engaging with other professionals' content. Aim for 5–10 meaningful interactions per day beyond your own posting.	Consistent presence in industry conversations that builds relationships and establishes you as an active, valuable community member
Performance and Reputation Monitoring	Set up tracking for your X analytics, monitoring engagement rates, profile visits, and follower growth. Regularly review your content for professional consistency and reputation impact.	Data-driven understanding of what content and engagement strategies generate the best professional outcomes and business opportunities

EIGHTEEN

Visual Storytelling on Instagram

Instagram's visual-first platform requires a fundamentally different approach to personal branding than text-based platforms. Success on Instagram means thinking like a visual storyteller who uses images, videos, and design elements to communicate expertise, personality, and professional values in ways that create emotional connections with your audience.

The most effective professional Instagram brands balance aspirational lifestyle content with authentic behind-the-scenes glimpses that humanize expertise while maintaining credibility. Your visual choices become extensions of your professional brand, communicating values, personality, and quality standards before followers read a single caption.

Visual storytelling on Instagram isn't about perfection—it's about consistency, authenticity, and strategic curation that supports your professional objectives while creating content that people genuinely want to follow and engage with regularly.

Developing Your Visual Brand Identity

Your Instagram visual brand should feel cohesive and intentional while remaining authentic to your actual lifestyle and professional context. This means developing consistent approaches to photography, color schemes, and content themes that reflect both your expertise and personality.

Color consistency creates instant recognition and professional polish. This doesn't mean every image must use identical colors, but establishing a palette of complementary colors that appear regularly throughout your content creates visual cohesion that makes your posts immediately recognizable.

Photography style consistency includes lighting preferences, composition approaches, and editing techniques that create a unified look across all your content. Whether you prefer bright, clean aesthetics or moody, dramatic lighting, consistency in your approach builds brand recognition.

Content theme balance involves mixing professional expertise with personal elements that make you relatable and interesting to follow. Business coaches might share workspace organization alongside family moments that demonstrate work-life integration, while consultants might combine client meeting insights with travel content that shows their mobile lifestyle.

Establish visual guidelines for yourself including preferred photo editing approaches, color schemes, and composition styles that you can reference when creating content. This ensures consistency while allowing creative flexibility within your established parameters.

Professional Lifestyle Integration

Instagram audiences want to see the person behind the professional brand, which requires thoughtful integration of personal and professional content that supports rather than undermines your credibility and expertise.

Behind-the-scenes content that shows your work process, preparation routines, or professional environment helps followers understand how you operate while maintaining appropriate professional boundaries. This might include workspace setups, preparation for client meetings, or the resources you use for professional development.

Personal elements that reinforce your professional values create authentic connections without compromising credibility. If you help busy professionals achieve better work-life balance, sharing content about your own family time, exercise routines, or relaxation practices demonstrates authentic commitment to the principles you teach.

Travel and location content can showcase the freedom and lifestyle that professional success enables while reinforcing your expertise. Remote work setups, client meetings in interesting locations, or speaking engagements in different cities all reinforce professional credibility while providing engaging visual content.

Essential Instagram Visual Brand Elements:

- **Consistent Filter/Editing Style**: Develop a signature photo editing approach using specific apps or presets that create unified visual aesthetics. Whether you prefer bright and airy, warm and cozy, or bold and dramatic, consistency in editing creates instant brand recognition.
- **Template Design System**: Create branded templates for quote graphics, educational carousels, announcement posts, and story highlights that maintain visual consistency while allowing efficient content creation. Use consistent fonts, colors, and layouts across all designed elements.
- **Grid Planning Strategy**: Plan your Instagram grid layout to create visual flow and balance between different content types. Use planning apps to preview how posts will look together and ensure your profile creates a strong first impression for new visitors.

- **Story Highlight Covers**: Design custom covers for your Instagram story highlights that match your brand aesthetic and clearly communicate the content themes. These covers become part of your profile's visual identity and help organize your content for easy navigation.
- **Professional Photography Standards**: Maintain consistent quality standards for all photos including proper lighting, sharp focus, and thoughtful composition. Invest in basic photography equipment or apps that help you create professional-looking content consistently.
- **Brand Color Integration**: Incorporate your brand colors naturally throughout your content through props, backgrounds, clothing choices, or graphic elements. This subtle integration reinforces brand recognition without appearing forced or unnatural.

REELS, STORIES, AND CAROUSELS THAT BUILD TRUST AND RELATABILITY

Instagram's diverse content formats provide multiple opportunities to showcase different aspects of your expertise and personality. Reels capture attention through entertaining or educational short-form video content, Stories build daily connection through casual, temporary content, and carousels allow for comprehensive education through multi-slide posts.

Each format serves different purposes in your overall brand strategy and appeals to different content consumption preferences among your followers. Mastering all three formats creates a comprehensive Instagram presence that engages various audience segments while reinforcing your professional brand through different content experiences.

The key to success with Instagram content formats is understanding how each one contributes to your larger professional objectives while providing genuine value that keeps followers engaged and attracts new audience members.

Instagram Reels for Professional Engagement

Reels provide powerful opportunities to demonstrate expertise through quick tutorials, industry insights, or behind-the-scenes professional content that entertains while educating your audience about your field.

Educational Reels that teach specific skills or share valuable insights often perform well because they provide immediate practical value. Break complex professional concepts into simple, actionable steps that viewers can implement immediately.

Behind-the-scenes Reels showing your work process, daily routines, or professional preparation create authentic connections while demonstrating your expertise in action. These might include client meeting preparation, content creation processes, or professional development activities.

Trend participation adapted to your professional context can increase visibility while maintaining brand relevance. Use trending audio or visual effects to create content that feels current and engaging while showcasing your expertise.

Industry commentary Reels that provide your perspective on current events, trends, or developments in your field position you as a thought leader while creating content that encourages discussion and engagement.

Stories for Daily Connection and Engagement

Instagram Stories provide opportunities for more casual, authentic content that builds personal connections with your audience through daily glimpses into your professional and personal life.

Professional updates about client successes, project completions, or business milestones shared through Stories feel more personal and authentic than formal posts while still reinforcing your expertise and success.

Q&A sessions using Instagram's question feature allow direct interaction with your audience while demonstrating your knowledge and willingness to help others with their challenges.

Behind-the-scenes Story content including preparation for client meetings, workspace organization, or professional development activities provides authentic glimpses into your work life without requiring polished presentation.

Real-time industry commentary on news, trends, or events relevant to your field shows that you stay current with industry developments and have valuable perspectives to share.

Carousel Posts for Comprehensive Education

Carousel posts allow for detailed educational content that demonstrates expertise while providing significant value to your audience. These multi-slide posts work well for step-by-step processes, comprehensive guides, or detailed case studies.

Educational carousels that break complex topics into digestible slides perform well because they provide substantial value while being easy to save and reference later. Focus on actionable insights that readers can implement immediately.

Case study carousels showcasing client results or project outcomes provide social proof while demonstrating your methodology and expertise in action. Include specific results and lessons learned that others can apply to their situations.

List-based carousels that compile resources, tools, tips, or strategies related to your expertise area provide immediate practical value while positioning you as a knowledgeable resource in your field.

Quote and insight carousels that present your professional philosophy, key principles, or important lessons learned help establish thought leadership while creating highly shareable content.

CONVERTING FOLLOWERS INTO LOYAL FANS OR CLIENTS

Instagram's visual nature creates strong emotional connections that can translate into business relationships when nurtured strategically through valuable content, authentic engagement, and clear communication about how you help your audience achieve their goals.

Converting Instagram followers into clients requires building trust through consistent value delivery, demonstrating expertise through helpful content, and creating natural pathways for interested followers to learn more about your services and potentially engage professionally.

The conversion process on Instagram typically happens gradually through relationship building rather than direct sales approaches. Your content should consistently provide value while subtly demonstrating your expertise and the results you help others achieve.

Building Trust Through Consistent Value

Establish yourself as a reliable source of valuable information by consistently sharing insights, tips, strategies, and resources that help your audience achieve their professional or personal goals.

Educational content that solves real problems for your audience builds trust while demonstrating your expertise and ability to help others achieve better outcomes in areas where you specialize.

Behind-the-scenes content that shows your work process, client interactions (with appropriate permissions), or professional development activities builds authenticity and trust by showing the real person behind the professional brand.

Client success stories and testimonials shared through various content formats provide social proof while demonstrating your ability to generate results for others facing similar challenges.

Consistent posting schedules and reliable content quality create expectations that build trust over time. Followers who know they can count on you for valuable content are more likely to consider your professional services when needs arise.

Strategic Call-to-Actions and Lead Generation

Use Instagram's link-in-bio feature to direct interested followers to lead magnets, consultation bookings, or comprehensive information about your services. Update this link regularly to align with current business objectives.

Story features like link stickers, swipe-ups (for accounts with 10k+ followers), and call-to-action buttons provide opportunities to direct engaged viewers to specific actions that support your business goals.

Direct message conversations often provide the most natural pathway from social media follower to business inquiry. Encourage DMs through your content and respond promptly with helpful information and clear next steps.

Email list building through valuable lead magnets promoted on Instagram creates owned marketing channels that allow deeper relationship building and more direct communication about your services.

Nurturing Long-term Relationships

Respond personally to comments and direct messages to build individual relationships with engaged followers who might become clients or valuable referral sources over time.

Share user-generated content and highlight follower successes to create community feeling while demonstrating the positive impact of your expertise and content.

Use Instagram's various features including Stories, Live videos, and IGTV to create different types of engagement opportunities that appeal to different followers' preferences and deepen relationships over time.

Track which content types and engagement strategies lead to the most business inquiries and focus your efforts on activities that generate qualified prospects rather than just general engagement.

Action Steps for Chapter 6:

Action Item	Task Description	Expected Outcome
Visual Brand Development	Establish your Instagram visual identity including color palette, photography style, and editing approach. Create templates for common content types and plan your grid layout strategy for visual consistency.	Cohesive Instagram brand that creates immediate recognition and professional credibility through consistent visual elements
Content Format Mastery	Create 10 pieces of content using different Instagram formats: 3 educational Reels, 3 value-packed carousels, daily Stories for one week, and regular feed posts. Analyze performance to identify most effective formats.	Expertise in all Instagram content formats with data-driven understanding of what resonates most with your specific audience
Engagement and Community Building	Develop a daily engagement routine including responding to comments, engaging with followers' content, and using Stories features for direct interaction. Aim for authentic connection rather than just promotional activity.	Strong community relationships that build trust and increase likelihood of converting followers into clients or advocates

Action Item	Task Description	Expected Outcome
Lead Generation System	Set up Instagram lead generation including optimized bio with clear value proposition, link-in-bio strategy, lead magnets promoted through Stories and posts, and systems for capturing and nurturing interested prospects.	Systematic approach to converting Instagram engagement into business inquiries and qualified prospects
Content Performance Analysis	Track Instagram analytics to identify top-performing content types, optimal posting times, and engagement patterns. Use insights to refine content strategy and focus on activities that drive both engagement and business results.	Data-driven Instagram strategy that maximizes both audience growth and business development opportunities

NINETEEN

Long-Form Influence on YouTube

YouTube represents the world's second-largest search engine and serves as the premier platform for establishing long-form thought leadership and comprehensive expertise demonstration. Unlike other social platforms that prioritize brief interactions, YouTube rewards depth, teaching ability, and sustained value delivery that builds lasting authority in your professional field.

Your YouTube channel functions as your digital university, providing space to share comprehensive knowledge, detailed case studies, and thorough explanations that establish you as the definitive expert in your niche. This depth of content creates stronger authority than any other platform while building an archive of valuable resources that attract new audiences for years after publication.

Building authority on YouTube requires strategic channel positioning that clearly communicates your unique expertise while creating content systems that consistently deliver exceptional value to your target audience. Your channel should become the destination where professionals in your field come for the most comprehensive, reliable information available.

Channel Strategy and Positioning

Define your channel's specific focus and value proposition with precision that differentiates you from the thousands of other professionals creating content in your general field. Rather than being "a marketing expert," position yourself as "the authority on email marketing for B2B SaaS companies" or "the go-to resource for productivity systems for creative entrepreneurs."

Your channel trailer and about section should immediately communicate who you serve, what specific problems you solve, and what unique perspective or methodology you bring to your field. New visitors should understand within seconds whether your channel is relevant to their needs and interests.

Develop content pillars that establish your expertise across the key areas where your target audience needs guidance. These pillars become recurring themes that demonstrate comprehensive knowledge while providing systematic coverage of your field.

Create a consistent publishing schedule that builds audience expectations and algorithmic favor. Whether you publish weekly or monthly, consistency in timing and quality builds trust while supporting YouTube's algorithm preferences for reliable content creators.

Authority-Building Content Types

Comprehensive tutorials and educational series that teach complex skills or processes in detail demonstrate your expertise while providing significant value that audiences bookmark and return to repeatedly. These in-depth resources become your channel's foundation content.

Case study analyses that break down successful projects, campaigns, or strategies provide behind-the-scenes insights that most professionals never share publicly. This transparency builds trust while demonstrating your analytical capabilities and real-world experience.

Industry analysis and commentary videos that interpret trends, developments, or news relevant to your field position you as someone who understands the broader context and can help others navigate change and opportunity.

Q&A sessions and response videos that address audience questions demonstrate responsiveness while covering topics that your audience specifically wants to learn about, ensuring content relevance and engagement.

YouTube Authority Development Framework:

1. **Expertise Documentation Strategy**: Create comprehensive content that documents your methodologies, processes, and frameworks in detail. This might include step-by-step guides for your signature approaches, detailed breakdowns of successful projects, or comprehensive explanations of industry concepts that establish your knowledge depth.
2. **Thought Leadership Positioning**: Develop unique perspectives, frameworks, or methodologies that become associated with your personal brand. Share contrarian viewpoints supported by evidence, create new terminology or concepts that advance your field, or synthesize existing knowledge in novel ways that provide fresh insights.
3. **Teaching Ability Demonstration**: Focus on creating content that actually helps viewers achieve specific outcomes rather than just sharing information. The ability to teach effectively—to transform complex concepts into actionable knowledge—establishes deeper authority than simply having expertise.
4. **Consistency and Reliability Building**: Maintain regular publishing schedules and consistent quality standards that build audience trust and algorithmic support. Establish yourself as someone your audience can count on for reliable, valuable content that consistently meets their professional development needs.

BALANCING ENTERTAINMENT WITH EDUCATION

Successful YouTube content combines educational value with engaging presentation that holds viewer attention throughout longer-form content. Unlike text-based platforms where information alone might suffice, YouTube's video format requires understanding of pacing, storytelling, and visual engagement that keeps audiences watching until completion.

The most successful professional YouTube channels understand that education and entertainment aren't opposing forces—they're complementary elements that work together to create memorable, impactful content that audiences eagerly anticipate and readily share with colleagues and networks.

Balancing these elements requires understanding your audience's learning preferences while developing presentation skills that make complex information accessible and engaging for video consumption.

Educational Content Structure

Structure educational videos with clear learning objectives stated upfront so viewers understand the specific value they'll gain by watching to completion. This expectation setting improves retention while helping you deliver on promised outcomes.

Break complex topics into logical segments with clear transitions that help viewers follow your reasoning and retain information effectively. Use verbal signposts like "first," "next," and "finally" to guide viewers through your content structure.

Include practical examples, case studies, or demonstrations that illustrate abstract concepts with concrete applications viewers can relate to their own situations and challenges.

Provide actionable takeaways and next steps that viewers can implement immediately after watching your video. This immediate applicability increases content value while encouraging viewers to return for additional guidance.

Entertainment Elements That Enhance Learning

Storytelling techniques, including personal anecdotes, client experiences, or case study narratives, make educational content more memorable while demonstrating real-world application of the concepts you're teaching.

Visual variety through screen sharing, graphics, props, or location changes maintains viewer interest during longer videos while supporting different learning styles and information processing preferences.

Personality integration through humor, enthusiasm, or authentic reactions to your content creates emotional connection while making educational material more enjoyable to consume.

Pacing variation that includes faster-paced segments for energy and slower explanations for complex concepts keeps viewers engaged while ensuring comprehension of important information.

Audience Engagement Optimization

Interactive elements like questions, polls, or calls-to-action encourage active participation rather than passive consumption, increasing engagement while providing valuable feedback about audience interests and needs.

Community building through consistent interaction with comments, addressing viewer questions in future videos, and acknowledging regular viewers by name creates loyalty while encouraging continued engagement.

Series development that creates anticipation for upcoming content keeps audiences returning while allowing comprehensive coverage of complex topics that require multiple videos to address completely.

Feedback integration that responds to viewer suggestions, questions, and requests demonstrates responsiveness while ensuring your content remains relevant to audience needs and interests.

MONETIZATION THROUGH ADS, SPONSORSHIPS, AND PRODUCTS

YouTube's mature monetization ecosystem provides multiple revenue streams that can generate significant income for channels that build engaged audiences and maintain consistent content creation. Understanding these monetization options allows strategic content creation that optimizes for both audience value and revenue generation.

Successful YouTube monetization requires building sufficient audience size and engagement to qualify for various programs while creating content that naturally integrates promotional elements without compromising educational value or audience trust.

The most sustainable YouTube revenue comes from diversified income streams that reduce dependence on any single monetization method while maximizing the value extracted from your content creation efforts and audience relationships.

YouTube Partner Program Optimization

YouTube's Partner Program provides ad revenue sharing that requires 1,000 subscribers and 4,000 watch hours in the past 12 months. Meeting these thresholds requires strategic content creation focused on audience building and engagement optimization.

Ad revenue optimization involves understanding YouTube's advertiser-friendly content guidelines while creating videos that attract premium advertising rates through professional presentation and valuable content that appeals to business audiences.

Video length strategy affects ad placement opportunities—videos longer than 8 minutes can include mid-roll ads that significantly increase revenue potential compared to shorter videos limited to pre-roll advertisements.

Audience demographics influence ad rates because advertisers pay premium prices for specific audience segments. Business and professional development content often attracts higher advertising rates than entertainment-focused channels.

Sponsorship and Brand Partnership Development

Direct sponsorships often provide more revenue than ad sharing while allowing greater control over promotional content integration and brand alignment with your professional values and audience interests.

Media kit development that showcases your audience demographics, engagement statistics, and previous collaboration examples helps attract brand partnerships while demonstrating professionalism that justifies premium sponsorship rates.

Integration strategies that naturally incorporate sponsored content into valuable educational material maintain audience trust while fulfilling sponsor objectives for authentic product or service endorsement.

Long-term partnership cultivation with brands that align with your audience's needs and interests creates ongoing revenue streams while reducing the effort required to continuously seek new sponsorship opportunities.

Product and Service Promotion

YouTube serves as powerful marketing for your own products and services through demonstration, education, and trust building that naturally leads to business inquiries and client relationships.

Course creation and promotion through YouTube content that provides sample teaching and demonstrates your expertise can generate significant revenue while serving audiences who want comprehensive education.

Consulting and coaching promotion through case studies, methodology explanations, and client success stories builds credibility while attracting high-value professional service inquiries.

Affiliate marketing integration for tools, resources, or services you genuinely use and recommend provides additional revenue while helping your audience access valuable resources relevant to your educational content.

Revenue Stream Integration Strategy

Diversified monetization approaches reduce risk while maximizing income potential from your content creation efforts. Combine ad revenue, sponsorships, product sales, and service promotion strategically.

Content calendar planning that balances educational value with monetization opportunities ensures audience trust while optimizing revenue generation through strategic promotional integration.

Analytics tracking that monitors which monetization methods generate the best results allows optimization of your revenue strategy while maintaining content quality and audience satisfaction.

Long-term strategy development that considers how different revenue streams support business growth while building sustainable income that justifies continued content creation investment.

Action Steps for Chapter 7:

Action Item	Task Description	Expected Outcome
YouTube Channel Setup and Optimization	Create or optimize your YouTube channel with professional branding, clear value proposition, compelling channel trailer, and organized playlists. Develop content pillars that establish your expertise across key topic areas.	Professional YouTube presence positioned as the authority hub for your specific expertise area with clear value communication
Content Creation and Publishing System	Plan and create your first 10 YouTube videos using educational content structure with entertainment elements. Establish a consistent publishing schedule and develop efficient production workflows.	Systematic approach to creating engaging, valuable content that builds authority while maintaining sustainable production processes
Audience Engagement Strategy	Implement community building practices including responding to comments, creating interactive content, asking for viewer feedback, and developing series that encourage return viewership.	Engaged YouTube community that provides feedback, shares content, and becomes a source of business opportunities and referrals
Monetization Planning	Research and plan your YouTube monetization strategy including Partner Program preparation, potential sponsorship targets, and integration of your own products or services. Create a media kit for brand partnerships.	Clear monetization roadmap that maximizes revenue potential while maintaining audience trust and content quality
Performance Analytics and Optimization	Set up YouTube analytics tracking for key metrics including watch time, subscriber growth, engagement rates, and revenue performance. Create a monthly review process for strategy optimization.	Data-driven approach to YouTube growth that identifies successful content types and monetization strategies for continued optimization

TWENTY

Community and Trust on Facebook

Facebook's mature platform provides unique opportunities for building meaningful professional communities through Groups and establishing business credibility through Pages. Unlike newer social platforms focused on discovery and viral content, Facebook excels at fostering ongoing relationships and facilitating deep community engagement around shared interests and professional development.

Facebook Groups create intimate spaces where professionals can share challenges, seek advice, and build relationships that often translate into business opportunities, referrals, and long-term professional collaborations. The platform's sophisticated group management tools allow for moderated discussions, exclusive content, and community building that develops genuine trust and mutual support.

Building successful Facebook communities requires understanding the platform's emphasis on meaningful social interaction and authentic relationship development rather than broadcasting or promotional activity. The most valuable Facebook presence focuses on serving others and facilitating valuable connections rather than self-promotion.

Strategic Group Creation and Management

Creating a Facebook Group around your area of expertise establishes you as a community leader while providing a platform for ongoing relationship building with potential clients, collaborators, and referral sources. Successful groups focus on specific niches rather than broad professional categories.

Choose group topics that address genuine ongoing needs in your professional community rather than topics that generate initial interest but lack sustainable engagement. "Marketing for Local Service Businesses" provides more focused value than "Business Marketing Tips."

Establish clear group guidelines that encourage helpful discussion while maintaining professional standards and preventing spam or inappropriate content. Well-managed groups attract higher-quality members who become valuable network connections.

Facilitate regular engagement through discussion prompts, expert interviews, resource sharing, and member spotlights that create value for participants while positioning you as a knowledgeable facilitator and connector.

Facebook Page Professional Development

Facebook Pages serve as your official business presence on the platform, providing credibility and comprehensive information about your professional services while integrating with Facebook's advertising and messaging systems.

Optimize your Page with complete business information, professional photography, clear service descriptions, and regular content that demonstrates expertise and builds trust with potential clients who research your background.

Use Page features like Services sections, Reviews, and Contact information to make it easy for potential clients to understand your offerings and initiate business conversations.

Maintain consistent posting schedules and professional presentation that reinforces your credibility while providing value to followers who use your Page to stay updated on your professional activities.

Community Engagement Best Practices:

- **Value-First Participation**: Focus on helping others solve problems, sharing valuable resources, and facilitating connections rather than promoting your services directly. Consistent value delivery builds trust that naturally leads to business opportunities.
- **Expert Commentary**: Provide thoughtful analysis and insights on industry developments, trending topics, and member questions that demonstrate your knowledge depth while helping others understand complex issues.
- **Resource Curation**: Share valuable tools, articles, studies, and resources that benefit group members while positioning yourself as someone who stays current with industry developments and helpful resources.
- **Connection Facilitation**: Introduce group members to each other when you identify potential collaboration opportunities, helping build community relationships while establishing yourself as a valuable network connector.
- **Content Moderation**: Maintain group quality by enforcing guidelines consistently, removing spam or inappropriate content, and guiding discussions toward productive outcomes that serve member needs.
- **Member Recognition**: Highlight member achievements, successful implementations of shared advice, and valuable contributions to create positive community culture and encourage ongoing participation.
- **Educational Series**: Create ongoing educational content series within groups that demonstrate expertise while providing systematic value that keeps members engaged and returning regularly.

- **Live Interaction**: Use Facebook Live features to host Q&A sessions, educational workshops, or informal discussions that create real-time connection and demonstrate personality alongside professional expertise.

LIVE SESSIONS FOR RELATIONSHIP BUILDING

Facebook Live provides powerful opportunities for real-time connection and authentic relationship building that creates stronger professional bonds than text-based interaction alone. Live video allows your audience to experience your personality, expertise, and communication style in ways that build trust and encourage business relationships.

The platform's Live features include audience interaction capabilities that create genuine two-way communication during broadcasts. This immediate feedback and personal recognition helps individual community members feel valued while building stronger overall group cohesion and engagement.

Successful Facebook Live sessions balance planned content with spontaneous interaction, creating valuable educational experiences while allowing personality and authenticity to build deeper professional relationships than polished, scripted content could achieve.

Planning Effective Live Sessions

Structure Live sessions with clear objectives and valuable content that justifies audience time investment while leaving space for spontaneous interaction and question response that creates personal connection.

Educational Live sessions that teach specific skills, share case studies, or provide industry analysis offer immediate value while demonstrating your expertise in real-time application and teaching ability.

Q&A format sessions allow direct interaction with your audience while addressing their specific challenges and questions, providing personalized value while showcasing your knowledge breadth and problem-solving approach.

Behind-the-scenes Live content that shares your work process, daily routines, or professional insights creates authenticity while demonstrating your expertise in natural, unscripted contexts.

Audience Interaction and Engagement

Acknowledge viewers by name during Live sessions to create personal connection while encouraging continued participation and return viewership for future broadcasts.

Respond to comments and questions in real-time to demonstrate accessibility and genuine interest in helping your audience, building trust that often translates into business inquiries.

Use Facebook's Live features like polls, reactions, and comment prompts to encourage active participation rather than passive viewing, creating more engaging experiences that build stronger community connections.

Follow up after Live sessions with participants who asked questions or showed strong engagement, using direct messages or additional content to continue conversations and build individual relationships.

Building Professional Authority Through Live Content

Demonstrate expertise through impromptu problem-solving during Live sessions, showing your ability to think on your feet while providing valuable guidance to real audience challenges.

Share case studies and real examples during Live sessions to illustrate concepts with concrete applications that help audiences understand how your expertise applies to their situations.

Address industry developments or current events during Live sessions to position yourself as someone who stays current with important trends while helping others understand implications and opportunities.

Create recurring Live session themes or series that build audience expectations while systematically covering important topics in your field, establishing comprehensive thought leadership over time.

USING FACEBOOK AS A FUNNEL FOR BUSINESS GROWTH

Facebook's sophisticated targeting capabilities and established user base make it an excellent platform for systematically building business funnels that move prospects from initial awareness through relationship development to client conversion. The platform's mature advertising system and organic reach opportunities work together to create comprehensive business development strategies.

Unlike newer platforms where business integration might feel forced, Facebook's users expect and accept business-oriented content when it provides genuine value. This acceptance allows for more direct business development approaches while maintaining focus on relationship building and value delivery.

Successful Facebook business funnels combine organic community building with strategic advertising to create systematic lead generation and client development processes that generate consistent business opportunities over time.

Lead Generation and List Building

Facebook's lead generation ads allow direct email capture within the platform, reducing friction while building email lists of prospects interested in your expertise and services.

Create valuable lead magnets specifically designed for Facebook audiences, considering the platform's demographics and user preferences for comprehensive resources and detailed information.

Use Facebook Groups to naturally promote lead magnets through helpful content that references additional resources available through email signup, creating organic lead generation within engaged communities.

Integrate lead generation across Facebook's various features including Page calls-to-action, Group pinned posts, and Live session promotions that direct interested prospects to your email capture systems.

Nurturing Prospects Through Value Delivery

Develop email sequences that continue the relationship building started on Facebook while gradually introducing your services and demonstrating your ability to help prospects achieve their goals.

Use Facebook's remarketing capabilities to stay visible to website visitors and email subscribers, creating multiple touchpoints that build familiarity and trust over extended periods.

Create exclusive content for Facebook communities that demonstrates additional value while building anticipation for your paid offerings and services.

Share client success stories and case studies through Facebook content that provides social proof while helping prospects understand the transformation possible through your services.

Converting Community Members to Clients

Identify the most engaged community members who consistently participate in discussions, share your content, and demonstrate genuine interest in your expertise as potential clients or referral sources.

Offer exclusive opportunities to community members such as discounted services, early access to new offerings, or complimentary consultations that reward community participation while generating business opportunities.

Use Facebook Messenger for direct communication with interested prospects, providing personalized attention while maintaining the informal, social context that makes business conversations feel natural.

Track which Facebook activities generate the most business inquiries and client conversions, focusing your effort on the community building and content strategies that produce the best business results.

Systematic Business Development Process

Create documented processes for moving Facebook connections through your business development funnel, from initial community engagement through client conversion and ongoing relationship maintenance.

Develop content calendars that balance community value with business objectives, ensuring consistent service to your community while systematically promoting your expertise and availability for professional services.

Use Facebook's analytics and advertising insights to understand which audience segments and content types generate the best business results, optimizing your funnel for maximum efficiency and conversion.

Establish systems for following up with prospects generated through Facebook, ensuring consistent communication that builds relationships while identifying qualified clients ready for your services.

Action Steps for Chapter 8:

Action Item	Task Description	Expected Outcome
Facebook Group Creation and Management	Create a focused Facebook Group around your expertise area with clear guidelines, valuable content themes, and regular engagement activities. Plan group growth strategy and community building activities.	Active professional community that positions you as a thought leader while providing networking and business development opportunities
Facebook Page Optimization	Set up or optimize your Facebook Page with comprehensive business information, professional branding, clear service descriptions, and regular valuable content that builds credibility with potential clients.	Professional Facebook presence that supports business credibility and provides easy contact methods for interested prospects
Live Session Strategy	Plan and execute weekly Facebook Live sessions including educational content, Q&A formats, and behind-the-scenes insights. Develop recurring themes that build audience expectations and expertise demonstration.	Regular live interaction that builds personal relationships and demonstrates expertise in real-time, natural contexts
Lead Generation System	Set up Facebook lead generation including compelling lead magnets, email capture systems, remarketing campaigns, and nurture sequences that move prospects toward client relationships.	Systematic approach to converting Facebook engagement into business leads and client relationships
Business Funnel Integration	Create a comprehensive business development process that leverages Facebook Groups, Page content, Live sessions, and advertising to systematically generate and nurture business opportunities.	Complete Facebook-based business development system that consistently generates qualified prospects and client conversions

TWENTY-ONE

TikTok for Fast-Track Growth

TikTok's algorithm represents the most democratic content distribution system among major social platforms, giving every creator, regardless of follower count, the opportunity to reach massive audiences through the "For You" page. Understanding how this algorithm operates becomes crucial for professionals seeking to build personal brands and establish expertise in an environment traditionally dominated by entertainment content.

The platform's algorithm prioritizes engagement velocity, content completion rates, and user interaction signals over follower counts or account age. This means that a well-crafted professional video from a new account can outperform content from established creators if it resonates strongly with its initial audience test group.

TikTok's algorithm operates on a feedback loop system that continuously learns from user behavior to serve increasingly relevant content. For professional creators, this means that consistent, valuable content aligned with your target audience's interests will eventually find and build the right community, even in TikTok's entertainment-heavy environment.

Understanding the For You Page Distribution

TikTok's For You page functions as the primary content discovery mechanism, exposing users to creators and topics beyond their direct followers. The algorithm tests every video with small audience segments before deciding whether to distribute content to larger audiences based on early engagement signals.

Initial performance metrics within the first few hours of posting determine whether the algorithm continues promoting your content. High completion rates, immediate comments, quick shares, and profile visits signal to the algorithm that your content provides value worthy of broader distribution.

The algorithm considers viewer behavior patterns, including how long someone watches your video, whether they watch it multiple times, whether they visit your profile after viewing, and whether they engage with your other content. These behavioral signals carry more weight than simple likes or follows.

Geographic and demographic matching plays a role in content distribution, with the algorithm initially showing your content to users with similar characteristics to your existing engaged audience. This creates opportunities for targeted professional content to reach relevant business audiences.

Engagement Signals That Drive Reach

Video completion rate serves as the strongest algorithmic signal, with videos watched to completion receiving significantly more distribution than content that loses viewers midway. Professional content must hook attention immediately and maintain interest throughout to succeed on the platform.

Comments generate more algorithmic value than likes because they require greater engagement investment from viewers. Creating content that naturally prompts questions, discussion, or response increases your content's distribution potential.

Shares and saves signal high-value content that users want to revisit or recommend to others. Professional content that provides actionable insights, valuable resources, or save-worthy tips often performs well due to these engagement behaviors.

Profile visits following video views indicate strong interest in your expertise and content, signaling to the algorithm that your video successfully introduced viewers to valuable creators they want to follow and engage with further.

Content Categories and Professional Positioning

TikTok's algorithm categorizes content into interest areas, making it essential for professional creators to establish clear content themes that help the algorithm understand your niche and target audience. Consistent posting within specific professional topics builds algorithmic recognition.

Educational content performs increasingly well on TikTok as the platform's audience matures and seeks valuable information alongside entertainment. Professional expertise shared through engaging formats can achieve significant reach while building genuine authority.

Business and career advice content has found strong audiences on TikTok, particularly among younger professionals seeking mentorship and career guidance. This creates opportunities for established professionals to share expertise while building personal brands.

Industry-specific content allows for niche audience building, with the algorithm becoming more effective at finding relevant viewers as it learns from engagement patterns within specific professional communities.

CREATING SHORT-FORM CONTENT WITH HIGH VIRALITY POTENTIAL

Professional content on TikTok succeeds when it adapts business expertise to the platform's fast-paced, engaging format while maintaining credibility and providing genuine value. The challenge lies in condensing complex professional insights into compelling 15-60 second videos that educate while entertaining.

Viral professional content typically combines trending formats with valuable expertise, using popular sounds, effects, or video structures to deliver business insights in ways that feel native to TikTok's culture while standing out for their practical value.

The most successful professional TikTok creators understand that virality comes from emotional resonance combined with practical value, creating content that people feel compelled to share because it either solved a problem or provided insights they want to pass along to their networks.

High-Performance Content Formats

"Day in the life" content that showcases professional routines, work environments, or business operations provides behind-the-scenes insights that build authenticity while demonstrating expertise in action. These videos humanize professional brands while providing valuable insights.

Quick tip or hack videos that solve specific problems in under 60 seconds perform well because they provide immediate, actionable value that viewers can implement right away. Focus on single concepts rather than comprehensive tutorials.

Reaction and commentary videos that provide professional perspectives on industry news, trends, or popular content allow for timely relevance while demonstrating expertise and unique viewpoints that add value to existing conversations.

Transformation or before-and-after content that shows results, improvements, or changes creates compelling visual narratives while demonstrating the effectiveness of your methods or approaches.

Viral Mechanics for Professional Content

Hook optimization becomes crucial in TikTok's fast-scroll environment, with the first 3 seconds determining whether viewers continue watching or move to the next video. Lead with your most compelling insight, surprising fact, or bold statement.

Pattern interrupts and visual variety maintain attention throughout short videos, using quick cuts, text overlays, props, or location changes to create dynamic content that holds viewer interest better than static talking-head formats.

Trending audio adaptation allows professional content to benefit from existing viral momentum while maintaining educational value. Use popular sounds as background for business tips or adapt trending audio to professional contexts.

Cliffhanger endings and series creation encourage viewers to follow your account for additional content, building ongoing audience relationships while allowing comprehensive coverage of complex topics across multiple videos.

Content Planning for Consistency and Growth

Batch content creation allows for consistent posting while maximizing time efficiency. Plan multiple videos around single themes, locations, or setups to create several pieces of content during focused creation sessions.

Trend monitoring and adaptation help professional creators participate in viral moments while maintaining brand relevance. Set up alerts for trending hashtags, sounds, and formats that could work with your area of expertise.

Content calendars that balance evergreen professional tips with timely industry commentary create consistent value while allowing for responsive content that capitalizes on current events and trending topics.

Performance analysis and iteration based on TikTok analytics helps identify which content types, posting times, and topics generate the best engagement with your specific audience, allowing for strategic optimization over time.

TURNING TRENDS INTO AUTHORITY-DRIVEN CONTENT

The key to professional success on TikTok lies in adapting viral trends and popular formats to showcase expertise while maintaining the entertainment value that drives platform engagement. This requires understanding how to participate in TikTok culture while positioning yourself as a credible authority in your field.

Successful trend adaptation involves identifying the core appeal of popular content formats and reimagining them through your professional lens, creating content that feels familiar to TikTok users while providing unique value through your expertise and perspective.

The most effective approach balances trend participation with original thought leadership, using trending formats as vehicles for sharing valuable professional insights rather than simply copying popular content without adding substantive value.

Trend Analysis and Professional Application

Monitor trending hashtags, sounds, and video formats to identify opportunities for professional adaptation that feel natural rather than forced. Not every trend will work with professional content, so selective participation maintains authenticity.

Analyze why certain trends become popular, whether through emotional appeal, practical value, entertainment factor, or shareability, then identify how these success factors can be applied to your professional content.

Time trend participation strategically to maximize reach while trends are still gaining momentum rather than participating after interest peaks. Early adoption often generates better results than late-cycle participation.

Maintain your professional brand voice and values while participating in trends, ensuring that trend adaptation enhances rather than conflicts with your desired professional positioning and expertise demonstration.

Content Adaptation Strategies

Educational trend adaptation transforms popular formats into learning opportunities by replacing entertainment elements with professional insights while maintaining the engaging structure that made the original trend successful.

Behind-the-scenes trend participation shows your professional life, work environment, or business operations using popular formats, providing authenticity while demonstrating expertise in natural contexts.

Industry-specific trend adaptation takes general concepts and applies them specifically to your professional field, creating niche content that serves your target audience while participating in broader TikTok conversations.

Contrarian or unique perspective trends that challenge popular opinions or provide alternative viewpoints can generate significant engagement while establishing thought leadership, provided the perspectives are well-reasoned and valuable.

Building Authority Through Strategic Trend Participation

Consistent value delivery through trend participation builds recognition as someone who provides useful professional insights regardless of format, establishing authority that transcends individual viral moments.

Expert commentary on trending topics related to your field positions you as someone who stays current with industry developments while having valuable perspectives to share about ongoing conversations.

Original trend creation based on your professional expertise can generate significant reach while establishing you as someone who contributes new ideas to TikTok culture rather than just adapting existing content.

Cross-platform promotion of successful TikTok content helps build authority across multiple channels while maximizing the reach and impact of your trend participation and professional insights.

Long-term Authority Building Strategy

Document your expertise systematically through trend participation, creating an archive of professional insights delivered through engaging formats that demonstrate both industry knowledge and communication skills.

Build recurring content themes that become associated with your professional brand, creating expectations among followers while establishing yourself as the go-to source for specific types of professional guidance.

Engage authentically with other professionals on the platform to build community relationships while expanding your network and creating opportunities for collaboration and cross-promotion.

Track which trend adaptations generate the most business inquiries, profile visits, and follower growth from your target professional audience, focusing efforts on trend types that best serve your business development goals.

Action Steps for Chapter 9:

Action Item	Task Description	Expected Outcome
TikTok Profile Optimization	Set up a professional TikTok profile with a clear bio describing your expertise, professional headshot, and links to other platforms. Develop content themes that balance professional value with platform culture.	Professional TikTok presence optimized for both algorithm success and business credibility building
Content Creation System	Develop a batch content creation process for producing 20–30 TikTok videos efficiently. Create templates for different content types including quick tips, behind-the-scenes, and trend adaptations.	Scalable system for consistent TikTok posting that maintains quality while maximizing time efficiency
Trend Monitoring and Adaptation	Set up systems for monitoring trending hashtags, sounds, and formats relevant to your industry. Create a process for quickly adapting trends to your professional expertise while maintaining authenticity.	Strategic approach to trend participation that builds authority while leveraging viral momentum for increased reach
Engagement and Community Building	Develop a daily engagement routine including responding to comments, engaging with other professionals' content, and participating in industry conversations. Focus on building genuine relationships within your professional niche.	Active TikTok community engagement that builds professional relationships and increases content visibility
Performance Tracking and Business Integration	Monitor TikTok analytics to identify best-performing content types and optimal posting strategies. Track how TikTok engagement translates to business opportunities and adjust strategy accordingly.	Data-driven TikTok strategy that maximizes both platform growth and business development outcomes

TWENTY-TWO

Content Strategy Across Platforms

Creating unique content for every social media platform quickly becomes overwhelming and unsustainable for busy professionals building personal brands. The most successful creators develop systematic approaches to content repurposing that maximize the value extracted from each piece of original content while adapting the format and messaging to meet the unique requirements and audience expectations of different platforms.

Effective content repurposing goes beyond simply posting identical content across multiple platforms. It requires understanding how the same core message can be adapted, reformatted, and recontextualized to provide optimal value and engagement on each platform while maintaining consistent brand messaging and expertise demonstration.

Strategic repurposing allows professionals to maintain an active presence across multiple platforms without proportionally increasing content creation time, creating efficiency that supports sustainable long-term personal brand building and business development.

The Content Multiplication Framework

Start with comprehensive, high-value content pieces that can be broken down into multiple smaller formats. A detailed case study can become a LinkedIn article, an Instagram carousel, a YouTube video, a Twitter thread, and several individual social media posts, each adapted for platform-specific audience preferences.

Identify the core insights, key points, and valuable takeaways from each piece of original content that can be extracted and reformatted for different consumption preferences and attention spans. This approach ensures each platform receives genuine value rather than diluted versions of comprehensive content.

Create content hierarchies that move from comprehensive formats (like YouTube videos or blog posts) down to quick highlights and key insights suitable for platforms with shorter attention spans. This top-down approach maintains quality while creating multiple touchpoints with your audience.

Develop template systems for common content adaptations that streamline the repurposing process. Templates for turning long-form content into social media carousels, video highlights, or quote graphics create efficiency while maintaining professional presentation standards.

Platform-Specific Adaptation Strategies

LinkedIn repurposing focuses on professional insights, industry analysis, and business applications of your content. Take personal anecdotes from other platforms and reframe them with emphasis on professional lessons learned and business implications.

Instagram adaptation emphasizes visual appeal and lifestyle integration, taking business concepts and presenting them through visually engaging formats that show personality alongside expertise. Behind-the-scenes content works particularly well for Instagram adaptation.

Twitter/X repurposing involves distilling complex ideas into concise, impactful statements that can spark discussion. Break comprehensive content into key insights, surprising statistics, or thought-provoking questions that encourage engagement.

TikTok adaptation requires transforming educational content into engaging, quick-consumption formats that provide immediate value. Focus on single concepts from longer content pieces and present them through trending formats or engaging visual approaches.

YouTube repurposing can expand brief social media content into comprehensive educational resources, using social media posts as research validation for topics that deserve deeper exploration and detailed explanation.

Facebook adaptation works well for community-building content that encourages discussion and relationship building around professional topics, taking individual insights and positioning them as conversation starters for your professional community.

Content Repurposing Systems and Workflows:

1. **Master Content Creation**: Develop comprehensive, high-value content pieces (blog posts, detailed case studies, in-depth tutorials) that serve as source material for multiple platform adaptations. This approach guarantees that all repurposed content is based on substantial value.
2. **Platform Priority Mapping**: Identify your primary and secondary platforms based on business goals and audience preferences, then create repurposing workflows that prioritize your most important platforms while ensuring quality across all channels.
3. **Format Template Library**: Create standardized templates for common content types across platforms including quote graphics, carousel layouts, video scripts, and text post structures that maintain brand consistency while enabling efficient adaptation.

4. **Content Calendar Integration**: Plan repurposing as part of your content calendar, scheduling platform-specific versions of core content to maximize reach while avoiding audience overlap that might see identical content across multiple platforms.
5. **Quality Control Systems**: Establish review processes that ensure repurposed content maintains quality standards and provides genuine value rather than appearing as lazy cross-posting that doesn't serve audience needs effectively.
6. **Performance Tracking**: Monitor which types of repurposed content perform best on each platform to optimize your adaptation strategies and focus efforts on the most effective approaches for your specific audience and business goals.

SCHEDULING, CONSISTENCY, AND AUTOMATION

Consistent content publishing across multiple platforms requires systematic approaches that maintain regular audience engagement without overwhelming your daily schedule or compromising content quality. The most successful personal brands develop automated systems that handle routine posting while preserving authentic engagement and timely responses to audience interaction.

Effective scheduling and automation strikes the balance between efficiency and authenticity, using technology to handle mechanical aspects of content distribution while maintaining the human touch that builds genuine relationships and trust with your professional audience.

Strategic automation focuses on content distribution rather than content creation or community engagement, ensuring that your personal brand maintains the authentic interaction that social media audiences value while creating sustainable systems for long-term brand building.

Content Scheduling Strategy

Optimal posting times vary by platform and audience demographics, requiring research into your specific audience's online behavior patterns. LinkedIn audiences typically engage during business hours, while Instagram users might be more active during evenings and weekends.

Platform-specific scheduling considers not only timing but also content type preferences at different times. Educational content might perform better during weekday business hours, while lifestyle or behind-the-scenes content could generate better engagement during personal time.

Frequency optimization involves finding the sweet spot between staying visible and avoiding audience fatigue. Different platforms have different optimal posting frequencies, with Twitter supporting multiple daily posts while LinkedIn might be more effective with weekly comprehensive content.

Global audience considerations become important as your personal brand grows internationally. Consider time zones of your key audience segments when scheduling content to ensure maximum visibility and engagement potential.

Automation Tools and Best Practices

Content scheduling platforms like Buffer, Hootsuite, Later, or Creator Studio allow batch content preparation and automated posting across multiple platforms while maintaining platform-specific optimization for each post.

Automation rules should include platform-specific hashtag strategies, optimal posting times, and content format adaptations that ensure each scheduled post feels native to its platform rather than obviously automated.

Engagement monitoring systems alert you to comments, messages, and mentions that require personal response, ensuring that automation doesn't create barriers to genuine community interaction and relationship building.

Content queue management involves maintaining scheduled content pipelines that allow for timely responses to current events, industry developments, or trending topics while ensuring consistent posting during busy periods.

Maintaining Authenticity Within Automated Systems

Personal engagement windows should be built into your schedule to respond to comments, participate in conversations, and engage with your community's content in ways that automation cannot replicate effectively.

Real-time content creation allows for spontaneous posts that respond to current events, industry news, or personal insights that couldn't be scheduled in advance, maintaining the authentic, immediate feel that builds trust.

Community interaction protocols ensure that automated systems don't interfere with genuine relationship building by clearly separating automated posting from personal engagement and relationship development activities.

Content variation within automated systems prevents your feed from feeling robotic by including different content types, tones, and formats that showcase various aspects of your expertise and personality.

ANALYTICS: HOW TO MEASURE GROWTH AND IMPACT

Understanding performance metrics across different platforms enables data-driven decisions about content strategy, platform prioritization, and resource allocation that optimize both audience growth and business development outcomes. Each platform provides unique analytics that reveal different aspects of audience behavior and content effectiveness.

Effective analytics goes beyond vanity metrics like followers and likes to focus on engagement quality, business impact, and audience actions that align with your professional objectives and personal brand goals. The most valuable metrics indicate audience interest, trust building, and business opportunity generation.

Cross-platform analytics comparison helps identify which platforms generate the best return on investment for your specific goals, allowing strategic resource allocation that maximizes impact while maintaining sustainable content creation practices.

Key Performance Indicators by Platform

LinkedIn analytics focus on professional engagement including profile views, connection requests, post engagement rates, and click-through rates to your website or business offerings. These metrics indicate growing professional recognition and business opportunity potential.

Instagram metrics emphasize visual engagement including Story completion rates, Save rates, Share rates, and profile visits following post engagement. Instagram analytics also provide demographic information that helps understand audience composition and preferences.

YouTube analytics offer comprehensive data including watch time, audience retention, click-through rates, and subscriber conversion rates that indicate content quality and teaching effectiveness. YouTube metrics also show traffic sources that reveal how audiences discover your content.

Twitter/X analytics track engagement rates, profile visits, link clicks, and mention quality that indicate thought leadership recognition and industry influence. Twitter metrics also show hashtag performance and conversation participation effectiveness.

TikTok analytics focus on completion rates, share rates, and profile visits that indicate content virality potential and audience building effectiveness within the platform's algorithm-driven environment.

Facebook analytics provide community engagement data including group activity, post reach, and audience demographics that help understand community building effectiveness and business lead generation potential.

Business Impact Measurement

Lead generation tracking connects social media activity to email signups, consultation requests, and other business inquiries that represent tangible professional opportunities resulting from personal brand building efforts.

Client acquisition attribution helps identify which platforms and content types generate the most business opportunities, allowing strategic focus on activities that best support professional growth and revenue generation.

Professional recognition metrics including speaking invitations, media mentions, collaboration requests, and industry award nominations indicate growing authority and influence within your professional field.

Network growth quality assessment examines not just follower numbers but the professional relevance and engagement quality of your growing network, focusing on connections that provide business value and collaborative potential.

Strategic Optimization Based on Analytics

Content performance analysis identifies which topics, formats, and posting strategies generate the best engagement and business results, allowing optimization of future content creation efforts for maximum impact.

Audience development insights reveal demographic trends, engagement patterns, and growth sources that inform strategic decisions about platform focus, content themes, and business development approaches.

ROI calculation for social media efforts involves comparing time and resource investment against business outcomes, including lead generation, client acquisition, and professional opportunity development, to ensure sustainable brand-building practices.

Platform prioritization based on performance data helps allocate limited time and resources to platforms and activities that generate the best results for your specific professional objectives and business goals.

Action Steps for Chapter 10:

Action Item	Task Description	Expected Outcome
Content Repurposing System Development	Create a systematic workflow for turning one piece of comprehensive content into 6–8 platform-specific adaptations. Develop templates and processes that maintain quality while maximizing efficiency.	Streamlined content creation process that maintains active presence across multiple platforms without proportional time investment
Scheduling and Automation Setup	Implement content scheduling tools and establish optimal posting times for each platform. Create automated workflows that handle distribution while preserving authentic engagement opportunities.	Consistent content publishing across platforms with systems that support sustainable long-term brand building

Action Item	Task Description	Expected Outcome
Analytics Dashboard Creation	Set up comprehensive analytics tracking across all platforms focusing on business impact metrics rather than vanity metrics. Create monthly reporting systems that inform strategic optimization decisions.	Data-driven understanding of which platforms and content types generate the best professional and business outcomes
Cross-Platform Content Calendar	Develop an integrated content calendar that coordinates messaging across platforms while adapting format and emphasis for each platform's unique audience and culture. Plan content themes and campaigns systematically.	Strategic approach to multi-platform content that maximizes reach while maintaining consistent brand messaging and professional positioning
Performance Optimization Protocol	Establish regular review cycles for analyzing performance data and optimizing content strategy based on engagement patterns, business lead generation, and professional opportunity development across all platforms.	Continuous improvement system that ensures social media efforts generate maximum business value and professional recognition

TWENTY-THREE

Monetizing Your Personal Brand

Personal brand monetization extends far beyond traditional employment or single-service business models to encompass multiple revenue streams that leverage your expertise, audience, and reputation in different ways. The most successful personal brands develop diversified income portfolios that reduce dependence on any single revenue source while maximizing the value extracted from brand building efforts.

Understanding various monetization options allows strategic development of income streams that align with your expertise, audience size, and business objectives while creating multiple pathways for professional and financial growth. Each income stream requires different approaches but benefits from the credibility and audience built through consistent personal brand development.

The key to sustainable personal brand monetization lies in balancing short-term revenue opportunities with long-term brand building activities that compound over time to create increasing value and income potential.

Sponsorship and Partnership Development

Brand partnerships provide revenue opportunities by connecting your audience with products, services, or companies that align with your expertise and values. Successful sponsorships feel authentic to your audience while providing genuine value rather than obvious promotional content.

Corporate partnerships can include content sponsorship, event partnerships, advisory roles, or brand ambassadorships that provide ongoing income while leveraging your professional credibility and audience influence for business value.

Product partnerships with companies whose offerings serve your audience create win-win relationships where your endorsement drives business results while your audience gains access to valuable resources and solutions.

Negotiating partnership terms requires understanding your audience demographics, engagement rates, and influence metrics while clearly communicating the value you provide to potential partners through authentic endorsement and professional credibility.

Speaking Engagement Opportunities

Professional speaking generates revenue while building authority and expanding your network through conference presentations, workshop facilitation, corporate training, and keynote addresses that showcase your expertise to new audiences.

Speaker fee structures vary based on event type, audience size, travel requirements, and your established reputation, with fees ranging from modest honorariums for local events to substantial fees for major conference keynotes.

Speaking topics should align with your established expertise while addressing current industry needs, challenges, or opportunities that conference organizers and corporate clients want their audiences to understand and address.

Building a speaking portfolio requires developing signature presentations, collecting video testimonials, creating professional speaker materials, and networking within conference and event planning communities.

Consulting and Professional Services

Consulting services leverage your expertise to help clients solve specific problems or achieve desired outcomes, typically commanding higher hourly rates than traditional employment while providing flexibility and diverse project opportunities.

Service packaging transforms consulting into scalable offerings like strategic planning intensives, implementation programs, or ongoing advisory relationships that provide predictable revenue while serving client needs comprehensively.

Client acquisition through personal brand activities creates inbound inquiry flow that reduces marketing costs while attracting clients who already understand your expertise and approach through your content and reputation.

Premium pricing becomes possible when your personal brand establishes clear differentiation, demonstrated results, and perceived value that justifies higher fees compared to less established competitors.

Digital Product Development

Online courses transform your expertise into scalable educational products that generate revenue without requiring proportional time investment for each sale, creating passive income potential that grows over time.

Digital resources like templates, frameworks, guides, or software tools provide immediate practical value while generating revenue from your systematized knowledge and proven methodologies.

Membership communities combine ongoing education, networking, and support into recurring revenue models that build long-term relationships while providing consistent income streams.

Book publishing, whether traditional or self-published, creates credibility assets that generate both direct revenue and indirect business opportunities through enhanced authority and visibility.

BUILDING CREDIBILITY TO ATTRACT PARTNERSHIPS

Partnership opportunities increase dramatically as your personal brand builds credibility through consistent value delivery, demonstrated expertise, and professional recognition within your industry. Companies and organizations prefer partnering with individuals who enhance their reputation while delivering authentic value to target audiences.

Building partnership-worthy credibility requires systematic development of trust indicators, including audience engagement, professional achievements, media recognition, and peer endorsements that validate your expertise and influence within your field.

The most valuable partnerships emerge from genuine relationship building rather than transactional networking, requiring long-term commitment to providing value and building trust with potential partners before commercial opportunities develop.

Establishing Professional Credibility

Thought leadership content that provides unique insights, challenges conventional thinking, or introduces innovative approaches positions you as someone worth partnering with for access to fresh perspectives and valuable expertise.

Media appearances, including interviews, article contributions, and expert commentary, build third-party validation while expanding your visibility beyond your immediate network and social media following.

Professional recognition through industry awards, certifications, speaking invitations, and peer acknowledgments provides external validation of your expertise and professional standing within your field.

Client results and case studies demonstrate your ability to generate real outcomes for others, providing potential partners with confidence that collaboration will produce measurable value rather than just theoretical expertise.

Building Strategic Relationships

Industry networking through conferences, professional associations, and peer groups creates opportunities to build relationships with potential partners while staying current with industry developments and collaboration opportunities.

Content collaboration with other professionals through joint articles, co-hosted events, or shared projects demonstrates your ability to work effectively with others while expanding your network and visibility.

Advisory roles and board positions provide opportunities to contribute expertise while building relationships with business leaders who might become collaboration partners or recommend you for partnership opportunities.

Referral relationship development with complementary professionals creates mutual support networks that generate business opportunities while providing partners with trusted resources for their clients and connections.

Demonstrating Partnership Value

Audience metrics and engagement data provide concrete evidence of your reach and influence that potential partners can use to justify collaboration investments and predict partnership outcomes.

Success story documentation from previous partnerships, client work, or collaborative projects provides evidence of your ability to deliver results and work effectively with others in professional relationships.

Professional reputation management through consistent quality delivery, reliable communication, and ethical business practices builds the trust foundation necessary for high-value partnership opportunities.

Unique value proposition articulation that clearly communicates what you bring to partnerships helps potential collaborators understand how working with you will enhance their objectives and provide competitive advantages.

CREATING A LONG-TERM BRAND MONETIZATION ROADMAP

Sustainable personal brand monetization requires strategic planning that balances immediate income needs with long-term wealth-building opportunities. The most successful personal brands develop systematic approaches that compound value over time rather than optimizing only for short-term revenue maximization.

A comprehensive monetization roadmap considers your expertise development, audience growth, platform expansion, and business evolution to create increasing value and income potential that builds sustainable professional and financial security.

Long-term thinking in personal brand monetization means making strategic decisions about partnerships, pricing, and opportunities that enhance rather than diminish your brand value while building assets that generate ongoing income beyond active work.

Monetization Stage Development

Early-stage monetization focuses on building credibility and establishing market positioning while generating modest income through services, small partnerships, or initial digital products that validate your value proposition.

Growth-stage opportunities include expanded service offerings, higher-value partnerships, speaking engagements, and scalable digital products that leverage your established audience and proven expertise for increased income potential.

Mature-stage monetization incorporates passive income streams, strategic investments, licensing opportunities, and advisory roles that generate significant income while requiring minimal ongoing time investment.

Legacy-stage planning considers how your personal brand and expertise can create lasting value through educational institutions, published works, or business assets that continue generating income and impact beyond your active involvement.

Revenue Stream Integration Strategy

Complementary income streams that support and enhance each other create more sustainable revenue than competing or conflicting monetization approaches. Speaking engagements that promote your consulting services or digital products that demonstrate your coaching methodology.

Seasonal planning that considers industry cycles, conference schedules, and business planning periods helps optimize timing for different monetization activities while ensuring consistent income throughout the year.

Investment in business infrastructure, including systems, team members, and capabilities that support multiple income streams simultaneously, increases efficiency while enabling growth that wouldn't be possible through individual effort alone.

Exit strategy consideration for various income streams allows flexibility to adapt to changing market conditions, personal preferences, or new opportunities while maintaining overall business stability and growth trajectory.

Long-term Value Building

Intellectual property development, including proprietary methodologies, trademarked processes, or unique frameworks, creates defensible business assets that command premium pricing while providing competitive differentiation.

Brand equity building through consistent quality, reliable delivery, and positive reputation creates intangible assets that support all monetization activities while increasing overall business value.

Network development and relationship building create ongoing opportunities for partnerships, referrals, and collaborative ventures that compound over time to generate increasing business value and income potential.

Knowledge and expertise expansion through continuous learning, skill development, and experience accumulation ensures that your monetization capabilities grow over time rather than becoming outdated or less valuable in changing markets.

Action Steps for Chapter 11:

Action Item	Task Description	Expected Outcome
Income Stream Assessment and Planning	Evaluate your current expertise and audience to identify 3–5 potential income streams. Research market demand, pricing structures, and requirements for each option. Create development timeline and resource requirements.	Strategic monetization plan that leverages your unique strengths while building multiple revenue sources

Action Item	Task Description	Expected Outcome
Credibility Building Initiative	Implement systematic credibility building including thought leadership content creation, media outreach, professional recognition pursuit, and case study documentation. Focus on third-party validation sources.	Enhanced professional credibility that attracts partnership opportunities and justifies premium pricing
Partnership Opportunity Development	Research potential brand partners, speaking opportunities, and collaboration prospects. Create outreach materials including media kit, speaker one-sheet, and partnership proposal templates.	Active pipeline of partnership opportunities that could generate significant revenue and brand enhancement
Digital Product Creation	Choose one digital product format (course, guide, template, or membership) and create a comprehensive offering including content development, pricing strategy, and launch plan.	Scalable income source that generates revenue without proportional time investment
Long-term Monetization Roadmap	Create a 3-year monetization plan that shows progression from current state to desired income levels through strategic development of multiple revenue streams. Include specific milestones and success metrics.	Clear roadmap for building sustainable, diversified income through strategic personal brand development

TWENTY-FOUR

Scaling and Sustaining Your Brand

As your personal brand matures, strategic collaborations and joint ventures provide powerful opportunities to accelerate growth, expand reach, and create value that exceeds what you could achieve independently. These partnerships leverage combined expertise, audiences, and resources to create outcomes that benefit all parties while serving larger communities more effectively.

Successful collaborations require careful partner selection, clear value exchange, and structured agreements that protect everyone's interests while maximizing collective impact. The most valuable partnerships combine complementary strengths rather than competing capabilities, creating synergy that enhances rather than dilutes individual brand positioning.

Strategic collaboration becomes essential for scaling beyond individual capacity constraints while maintaining brand quality and authenticity. These partnerships allow access to new audiences, enhanced credibility, and expanded capabilities without compromising the personal brand foundation you've built.

Strategic Partnership Selection

Identify potential partners whose audience overlaps with but doesn't duplicate your target market, creating opportunities for mutual audience expansion without cannibalizing existing relationships. Look for professionals whose expertise complements rather than competes with your specialization.

Evaluate potential partners' brand reputation, content quality, and audience engagement to ensure alignment with your standards and values. Poor partner choices can damage your brand reputation even when collaboration intentions are positive.

Consider geographic, demographic, or industry expansion opportunities through partnerships with professionals who serve similar audiences in different markets, creating growth opportunities that leverage your expertise in new contexts.

Assess partnership compatibility through smaller collaborative projects before committing to major joint ventures, allowing relationship evaluation and working dynamic assessment without significant risk or investment.

Joint Venture Structure Development

Define clear roles, responsibilities, and contribution expectations for each partner to prevent confusion and conflict while ensuring equitable value exchange that benefits all parties appropriately.

Establish intellectual property agreements that protect individual brand assets while creating shared ownership of collaborative content and resources developed through joint efforts.

Develop revenue-sharing structures that reflect each partner's contribution while incentivizing continued collaboration and mutual success rather than creating competitive tension or resentment.

Create communication protocols and decision-making processes that ensure smooth collaboration while respecting individual brand autonomy and maintaining positive working relationships throughout the partnership.

Collaboration Value Maximization

Cross-promotion strategies that introduce each partner to the other's audience should feel natural and valuable rather than forced or purely promotional, providing genuine value to all audiences involved.

Content co-creation that brings together different areas of expertise results in valuable resources that neither partner could create alone, showcasing their teamwork and adding more value.

Event and program partnerships that combine resources and audiences can create larger, more impactful offerings while sharing costs and risks associated with significant business development investments.

Working with partners helps you connect with larger professional groups and discover new opportunities that go beyond just the current collaboration.

DELEGATING AND OUTSOURCING BRAND MANAGEMENT

Scaling your personal brand requires transitioning from handling all activities personally to building team capabilities that maintain brand quality while expanding capacity for growth and business development. Strategic delegation preserves your personal involvement in high-value activities while ensuring consistent brand representation across all touchpoints.

Effective delegation requires developing clear brand guidelines, training team members thoroughly, and creating quality control systems that maintain your brand standards while allowing others to handle routine tasks and operational responsibilities.

The goal of delegation isn't to remove yourself from your brand but to focus your personal involvement on activities that uniquely require your expertise, personality, and decision-making while building systems that support sustainable growth.

Team Building for Brand Support

Virtual assistants can handle scheduling, basic customer service, research, and administrative tasks that support your brand activities without requiring deep expertise or personal brand representation.

Content creators, including writers, designers, and video editors can develop materials under your guidance, maintaining brand consistency while expanding content production capacity beyond your personal time constraints.

Social media managers can handle posting schedules, routine engagement, and community management while escalating important conversations and relationship-building opportunities to your personal attention.

Business development support, including lead qualification, client onboarding, and project coordination, allows you to focus on high-value client relationships and strategic business development while ensuring smooth operational execution.

Quality Control and Brand Consistency

Brand guidelines documentation should include voice characteristics, visual standards, messaging frameworks, and response templates that enable team members to represent your brand authentically and consistently.

Review processes for all external communications ensure that delegated work meets your quality standards and accurately represents your expertise and personality before reaching your audience.

Training programs for team members should include brand education, communication standards, and escalation procedures that ensure appropriate handling of various situations while maintaining brand integrity.

Regular performance evaluation and feedback systems help team members improve their brand representation while identifying areas where additional training or process refinement might enhance consistency and quality.

Maintaining Personal Brand Authenticity

Reserve high-visibility activities, including media interviews, speaking engagements, and important client relationships, for your personal involvement to maintain authentic connection with your audience and key stakeholders.

Personal content creation for the most important brand communications ensures that your unique voice and perspective remain central to your brand while delegated activities handle supportive and operational functions.

Relationship management with key clients, partners, and industry contacts should remain your personal responsibility to maintain the authentic connections that drive business opportunities and collaborative partnerships.

Strategic decision-making about brand direction, partnership opportunities, and business development should involve your direct input to ensure alignment with your goals and values while leveraging team support for implementation.

BUILDING A BRAND THAT THRIVES BEYOND TRENDS

Sustainable personal brands are built on timeless principles and enduring value rather than trending topics or temporary cultural moments. While trend participation can provide short-term visibility, long-term brand success requires developing expertise, reputation, and relationship foundations that remain valuable regardless of changing platforms or popular topics.

Creating trend-resistant brand foundations requires focusing on fundamental human needs, universal professional challenges, and timeless success principles that maintain relevance across changing business environments and technological developments.

The most enduring personal brands combine consistent core values with adaptive communication methods, maintaining authentic expertise and helpful perspective while evolving presentation formats to serve audiences effectively across different platforms and cultural contexts.

Timeless Value Creation

Focus on solving fundamental problems that persist across technological and cultural changes rather than addressing only current trends or temporary challenges that might become irrelevant quickly.

Develop expertise in principles and frameworks that apply broadly across industries and situations rather than specializing exclusively in tools or methods that might become obsolete with technological advancement.

Build a reputation based on character, integrity, and consistent value delivery rather than controversial positions or trend-based attention that might damage long-term credibility and relationship potential.

Create educational content that teaches thinking processes and decision-making frameworks rather than just providing answers to current questions, ensuring ongoing value as circumstances change.

Platform Independence Strategy

Develop owned media, including email lists, websites, and direct communication channels that don't depend on social media platforms for audience access and relationship maintenance.

Create content that can be adapted across multiple platforms and formats rather than developing platform-specific content that becomes worthless if platforms change or decline.

Build direct relationships with your audience through personal interaction, events, and communication that create loyalty beyond platform convenience or algorithmic exposure.

Maintain presence across multiple platforms while avoiding overdependence on any single channel for business development or audience relationship building.

Legacy Brand Development

Document your expertise and methodology in comprehensive formats, including books, courses, or structured programs that provide lasting value and establish permanent authority in your field.

Mentor and develop other professionals who can extend your influence and carry forward your expertise contributions while building network relationships that support long-term business sustainability.

Create intellectual property, including frameworks, processes, or systems that can provide ongoing value and income potential while establishing lasting contributions to your professional field.

Build institutional relationships with organizations, educational institutions, or industry groups that provide stability and ongoing influence opportunities beyond individual social media presence or trending popularity.

Evolution and Adaptation Strategy

Keep an eye on changes in the industry and technology that could impact what your audience needs, while still focusing on your main skills and the value you offer that go beyond just specific tools.

Continuously develop new skills and knowledge areas that complement your established expertise while expanding your ability to serve evolving audience needs and market opportunities.

Regular brand evaluation and strategic planning help find changes that need to be made while keeping the core positioning and value delivery the same, which helps build long-term relationships.

Succession planning considerations for your personal brand, including knowledge transfer, relationship maintenance, and asset preservation, ensure that brand value extends beyond your individual involvement and contributes lasting value to your professional field.

Action Steps for Chapter 12:

Action Item	Task Description	Expected Outcome
Strategic Partnership Development	Identify 5–10 potential collaboration partners whose expertise complements yours. Research their audience alignment and brand reputation. Develop partnership proposals for joint content, events, or business ventures.	Active collaboration opportunities that expand reach and capabilities while maintaining brand integrity
Team Building and Delegation System	Define roles that can be delegated while maintaining brand quality. Hire and train virtual assistants or team members for content creation, social media management, and administrative support. Create quality control processes.	Scalable brand management system that expands capacity while preserving authenticity and quality standards

Action Item	Task Description	Expected Outcome
Brand Guidelines and Training Materials	Create comprehensive brand guidelines including voice characteristics, visual standards, response templates, and quality control processes. Develop training materials for team members who represent your brand.	Systematic approach to maintaining brand consistency as you scale operations and delegate responsibilities
Platform Independence Strategy	Develop owned media channels including email list, website, and direct communication methods. Create content systems that work across multiple platforms. Reduce dependence on any single platform for audience access.	Sustainable brand foundation that thrives regardless of platform changes or algorithm updates
Legacy Planning and Documentation	Begin documenting your expertise in permanent formats including comprehensive guides, frameworks, or educational programs. Plan for long-term brand sustainability and knowledge transfer.	Enduring brand assets that provide lasting value and establish permanent authority in your professional field

Your Words Could Inspire the Next Great Personal Brand

At the heart of every act of generosity is a simple truth: when we share what we know, we help others grow.

Have you ever thought about how even a small piece of encouragement or guidance can change the course of someone else's life?

Picture a reader standing at the starting line—eager to build a personal brand but overwhelmed by the maze of **LinkedIn, X, Instagram, YouTube, Facebook, and TikTok**. They may have valuable skills or stories to share but lack the confidence or clarity to put themselves out there. Your reflections on *Personal Branding: A Complete System to Build and Monetize a Unique Brand Using Influencer Strategies and Step-by-Step Marketing Tools on LinkedIn, X, Instagram, YouTube, Facebook, and TikTok* could be the nudge that convinces them it's possible.

That's why I'd be so grateful if you would leave an **honest review**. It's more than feedback—it's part of a shared mission to help others discover their voice, grow their influence, and take bold steps toward turning passion into opportunity.

By sharing your thoughts, you:

- **Guide future readers** who wonder if this book can help them rise above the noise.
- **Support the vision** of helping more people build brands that feel authentic and meaningful.
- **Amplify impact**, because your words can reach far beyond a single review.

Thank you sincerely for taking the time to share your perspective. And if you know someone who dreams of creating a brand that opens doors and inspires others, I'd be honored if you passed this book along to them as well.

Conclusion

The rules of business have fundamentally changed. Gone are the days when hiding behind corporate logos and company brands was enough to build a successful career or business. Today, people buy from people, not faceless corporations. They follow individuals who share their values, solve their problems, and inspire their aspirations.

THE SHIFT FROM CORPORATE IDENTITY TO PERSONAL IDENTITY IN BUSINESS

Traditional business operated on institutional trust. Companies spent decades building brand recognition through advertising, public relations, and corporate messaging. Employees were expected to be interchangeable parts of the corporate machine, their individual personalities absorbed into company culture and brand guidelines.

This model has become obsolete in the digital age. Consumers now expect transparency, authenticity, and personal connection from the brands they support. They want to know who stands behind the products and services they purchase. They seek relationships, not just transactions.

Consider how business decisions are made today. When choosing between two consultants with similar credentials, clients almost always select the one whose personal brand resonates with their values and communication style. When investors evaluate startups, they are betting on the founders' personal brands as much as the business concept. When companies hire executives, they consider not only professional qualifications but also how that person's personal brand aligns with and enhances their organizational goals.

The most successful professionals understand that their personal brand is their competitive advantage. It is what differentiates them in crowded markets, attracts opportunities, and commands premium pricing. Your personal brand becomes your business moat, something competitors cannot replicate because it is uniquely yours.

WHY PERSONAL BRANDING IS ESSENTIAL IN THE DIGITAL AGE

Digital platforms have democratized influence and entrepreneurship in ways never seen before. Anyone with expertise, passion, and strategic thinking can build a global audience and generate significant income. However, this same accessibility has created intense competition for attention and trust.

Without a strong personal brand, you become invisible in the digital noise. Your expertise remains hidden, your potential clients cannot find you, and opportunities pass to those who have mastered the art of personal branding. The choice is not whether to build a personal brand, but whether to build one intentionally or allow others to define it for you.

Search engines and social media algorithms favor individuals who consistently share valuable content and engage authentically with their communities. They reward personal brands that demonstrate expertise, build relationships, and create genuine value for their audiences. These platforms have become merit-based systems where the most helpful, authentic, and consistent personal brands rise to the top.

Modern consumers conduct extensive research before making purchasing decisions. They examine your online presence, read your content, watch your videos, and evaluate your credibility before engaging with your services or products. Your personal brand serves as your twenty-four-hour sales representative, building trust and demonstrating value even when you are not directly involved in the conversation.

The global nature of digital business means your personal brand can attract opportunities from anywhere in the world. Geographic limitations no longer constrain professional growth when your expertise and personality can reach international audiences through strategic personal branding.

WHAT THIS BOOK PROVIDES: A COMPLETE SYSTEM FROM BUILDING TO MONETIZING

This book provides a comprehensive, step-by-step system for building and monetizing a personal brand across all major digital platforms. You will learn not just what to do, but how to do it efficiently and effectively, with specific strategies tailored to each platform's unique characteristics and audience expectations.

The system covers every aspect of personal branding: discovering your unique value proposition, crafting compelling messages, designing professional visual elements, building authority across multiple platforms, creating engaging content consistently, and converting your influence into sustainable income streams.

Each chapter builds upon the previous one, creating a cohesive strategy that compounds your efforts rather than scattering them across disconnected tactics. You will understand how content created for one platform can be adapted and repurposed for others, how different platforms serve different functions in your overall brand strategy, and how to create synergies that amplify your impact.

The strategies presented here are based on proven methods used by successful personal brands across various industries. They are practical, implementable, and designed to generate real results rather than just theoretical understanding. Every concept is accompanied by specific action steps that allow immediate implementation.

READER PROMISE: TRANSFORM VISIBILITY INTO INFLUENCE AND INFLUENCE INTO INCOME

By the end of this book, you will have a complete roadmap for transforming from hidden expert to visible authority to profitable influencer. This transformation does not happen overnight, but it follows predictable patterns that you can accelerate through strategic action and consistent execution.

You will understand how to identify and communicate your unique value in ways that resonate with your target audience. You will know how to create content that builds trust, demonstrates expertise, and naturally leads to business opportunities. You will have specific strategies for each major social platform and understand how to integrate them into a cohesive personal branding system.

Most importantly, you will have a clear path for monetizing your personal brand through multiple income streams including consulting, speaking, digital products, sponsorships, and partnership opportunities. You will understand how to price your expertise, attract premium clients, and build a sustainable business around your personal brand.

The journey from anonymity to influence requires commitment and consistent action, but the rewards extend far beyond financial success. A strong personal brand provides professional security, creative fulfillment, and the satisfaction of helping others while building something uniquely yours.

Your expertise has value. Your perspective matters. Your audience is waiting for someone exactly like you to help them achieve their goals. This book provides the systematic approach to connect your unique value with the people who need it most, creating mutual benefit that forms the foundation of sustainable business success.

> *"The best time to plant a tree was 20 years ago. The second best time is now."*
>
> – Chinese Proverb

Your personal brand empire starts today.

www.ingramcontent.com/pod-product-compliance
Lightning Source LLC
Chambersburg PA
CBHW071431230926
51846CB00001B/190
9781971071923